★ ★ ★ ★ ★ ★ ★ ★ ★ ★ ★ ★ ★

The Voice of America

★ ★ ★ ★ ★ ★ ★ ★ ★ ★ ★ ★ ★ ★

The Voice of America

Merni Ingrassia Fitzgerald

Illustrated with photographs

Dodd, Mead & Company ★ New York

This book is dedicated to my parents,
TONY AND BILLIE INGRASSIA

All photographs are used through the courtesy and with the permission of Voice of America. The illustrations of the program clock and the VOA hand signals are reprinted from the *Voice of America Handbook*.

The excerpt from the story "How Thor Found His Hammer" from *Norse Stories* by Hamilton Wright Mabie is reprinted by permission of Dodd, Mead & Company.

Copyright © 1987 by Merni Ingrassia Fitzgerald
All rights reserved. No part of this book may be reproduced in any form without permission in writing from the publisher. Distributed in Canada by McClelland and Stewart Limited, Toronto. Manufactured in the United States of America

1 2 3 4 5 6 7 8 9 10

Library of Congress Cataloging-in-Publication Data
Fitzgerald, Merni Ingrassia. The Voice of America.
Includes index. Summary: Describes the activities and purposes of Voice of America, the radio organization that broadcasts music, news, and special features to countries all over the world.
  1. Voice of America (Organization)—Juvenile literature. [1. Voice of America (Organization) 2. International broadcasting.] I. Title.
HE8698.F5   1987   384.54′53   87-5346
ISBN 0-396-08937-2

Contents

	Foreword by Richard W. Carlson, DIRECTOR, VOICE OF AMERICA	vii
1 •	Broadcasting to the World	1
2 •	The Yankee Doodle Station	12
3 •	Music	21
4 •	News and Features	36
5 •	Special English	51
6 •	Children's Programming	59
7 •	Focus on the Soviet Union	66
8 •	VOA Listeners	72
9 •	Broadcasters	84
10 •	Radio Marti	99
11 •	VOA and You	107
	VOA Languages	112
	VOA Timeline	114
	Index	116

The author wishes to acknowledge the gracious cooperation and assistance of current and former members of the VOA staff, especially Richard R. Dow. The author also wishes to acknowledge her friend, Sherrod Shim, for her constant encouragement and support.

Foreword

By RICHARD W. CARLSON
DIRECTOR, VOICE OF AMERICA

The world has changed a great deal since the Voice of America began broadcasting nearly forty-five years ago, but its primary mission and objective has never wavered: dissemination of the truth. Tens of millions of people around the world are denied access to truth by their own governments and leaders—and they depend on VOA to fill the information gap.

Communications technologies have rendered physical boundaries obsolete. A word spoken in Washington, D.C., can be heard in the Far East one-fourth of a second later. Information can travel as far and as fast as sound itself. Now, even citizens living in the most closed societies have access to the airwaves, and that simple fact is changing the way world leaders govern and how they interact with other nations.

For Americans it means, more than ever before, that

Richard W. Carlson, Director of Voice of America

VOA plays a critical role in our foreign policy efforts. In the course of a week, many more than 120 million people tune in to VOA to hear the truth.

America's founding fathers declared: "Let the facts be submitted to a candid world." The Voice of America does just that every day for its listeners on behalf of the American people.

"Truth remains the ultimate weapon in the arsenal of Democracy," President Ronald Reagan noted at a VOA ceremony in 1982. Indeed, it is the ultimate weapon because it is inviolable. There is no way to imprison a mind. VOA is the medium by which the truth becomes known to the world, and it will continue to be one of America's best and brightest representatives overseas.

This book makes a valuable contribution toward building an understanding by young Americans of the Voice and its mission, its history, its programs, and its personalities. Merni Fitzgerald has captured the unique spirit of VOA in an entertaining and absorbing narrative chockfull of anecdotes which really bring the VOA story to life. This book will show schoolchildren across the United States that government-run international radio can be a powerfully effective tool of public diplomacy, and that all around the globe there are friends of VOA who have become friends of America.

December, 1986

VOA headquarters in Washington, D.C.

Broadcasting to the World

Every day an American radio station in Washington, D.C., broadcasts music, news, and feature programs to the people in every other country in the world except the United States. Over 120 million people tune in for information about the United States, and accurate news reports. Over 1,300 hours of programming are broadcast every week in over 40 different languages. This radio station is the Voice of America (VOA), the global radio network of the United States Information Agency.

VOA headquarters are located only blocks away from the popular Air and Space Museum and the Capital in Washington, but very few Americans count VOA among the "must-see" tourist attractions to visit during their vacations to the nation's capital. Despite the 10,000 persons who take the VOA tour every year, most Americans don't know much about VOA.

When VOA staff members traveled to Texas in the specially equipped VOA Voyager van to tape news stories, they were asked if they were members of a new religious group. The distinctive red, white, and blue VOA logo was clearly visible on both sides of the van, yet people still wondered what it was and asked questions.

This is because VOA does not broadcast directly into the United States. If an American has a shortwave radio, he would be able to listen to some VOA programs, but Americans are not the intended audience. As a result, most Americans do not know that VOA exists or what it does.

The U.S. government did not have any international radio station before 1940. Other means were used to give information about the United States to the people of the world—establishing libraries in foreign countries, distributing U.S. publications, supplying articles to foreign newspapers. During the period just before the outbreak of World War II, it was felt that more needed to be done to respond to German and Italian propaganda. So the first international radio broadcasting by the U.S. government began. The office of the Coordinator of Inter-American Affairs (CIAA) undertook this job, under the leadership of Nelson A. Rockefeller (later governor of New York and Vice President of the United States). Existing private radio stations, including some owned by NBC and CBS, were used. All CIAA broadcasts were directed to Latin American countries.

President Franklin D. Roosevelt decided to expand the international radio program. He established the Coordinator of Information office (COI) in July, 1941. The Foreign Information Service (FIS) was part of this government agency. The FIS offices were located at 270 Madison Av-

Broadcasting to the World ★ 3

President Franklin D. Roosevelt expanded international radio broadcasting.

enue in New York City. This is where the Voice of America came into existence.

The first VOA broadcast was in the German language on February 24, 1942, from the FIS offices in New York to Germany. Pearl Harbor had just been bombed 79 days earlier on December 7, 1941. The Germans were broadcasting Nazi propaganda to the world. President Roosevelt wanted to counter the lies, and let the world know what America stood for, and what Americans believed.

That very first VOA broadcast was done by William Harlan Hale. Years later, Hale remembered that night: "... by a strange irony, it so happened that before the broadcast I had an invitation to dinner and a ballet and was dressed accordingly for the occasion ... it was the last ballet I saw during the war, but it was a strange affair to appear before the microphone speaking to the enemy in full dress . . . All we called ourselves at that time was 'Voices of America.' I remember the first words I used in the German language said, *'Wir bringen Ihnen Stimmen aus America'*—in other words, 'We bring you voices from America.' It was only a few months later that we were finally given the official title, the Voice of America."

In the beginning, VOA was not intended as an information agency. It was seen as another weapon in the war. Its aim was to counter the wartime propaganda, and the programs were slanted to achieve that goal. The purpose of VOA was mentioned on that first broadcast: "Daily at this time we shall speak to you about America and the war. The news may be good or bad. We shall tell you the truth."

The VOA broadcasters purposely spoke the foreign language with an American accent. This was done so that the broadcast would be recognized as an American program, and so that the Germans would not charge that German refugees were doing the broadcasts as revenge against their homeland. In some places VOA was known as the "Yankee Doodle Hour," because the opening and closing music was "Yankee Doodle." This music is still used today to open and close VOA programs.

In June, 1942, the VOA was placed in the Office of War Information (OWI), which combined several agencies, including FIS and COI. The CIAA continued to broadcast programs to Latin America. When the war ended, the future of VOA was in doubt. It had been formed during the war to broadcast propaganda to the people of the world. There had never been a peacetime government international radio program. VOA personnel was reduced, and some of the language services were dropped. But in August of 1945, President Harry Truman transferred the international radio functions (including VOA) from the OWI and CIAA to the State Department.

Construction of VOA transmitters, begun in 1942, had continued and by 1945 there were transmitters in Dixon and Delano, California, and in Bethany, Ohio. Today, VOA has more than one hundred radio transmitters located in the United States and twelve other countries, three hundred antenna groups, and communication satellites.

In 1946 and 1947, the staff of VOA had to fight for its very existence. There were budget problems, and there

A VOA transmitter relay station

was no legislation making the Voice of America permanent. World events helped the status of VOA considerably. In 1947 the "cold war" with the Soviet Union began. The Soviet Union radio broadcasts were beginning to criticize America and American policies. In response, VOA began a Russian Language Service. VOA was "rediscovered" to help in this East/West conflict. It was again seen as a means to counter propaganda, this time from the communists.

The Russians began jamming VOA broadcasts in February, 1948—using the same frequencies so that VOA programs could not be heard—but 1948 became an important year in the history of VOA. Until then, almost 75 percent of the VOA programming was handled by private broadcasters, including NBC and CBS. In October, 1948, VOA became directly responsible for all of the programming it broadcasts. A trend began during this period. Foreign Service officers were appointed to VOA staff positions. This helped VOA's relationship with the State Department. Today, Foreign Service officers continue to hold some of the top management jobs at VOA.

It was in 1948 that Congress passed a law making VOA a permanent peacetime overseas information program. The law was the United States Information and Educational Exchange Act of 1948, also known as the Smith-Mundt Act because it was introduced by Congressman Karl Mundt of South Dakota and cosponsored by Senator H. Alexander Smith of New Jersey.

The committee that reported this bill to the Senate noted that "truth can be a powerful weapon on behalf of peace . . . [this bill] will constitute an important step in the right direction toward the adequate dissemination of the truth about America; our ideals, and our people."

Broadcasting to the World ★ 7

H. Alexander Smith of New Jersey, left, and Karl Mundt of South Dakota cosponsored the United States Information and Educational Exchange Act. Here they are interviewed on VOA.

The Smith-Mundt Act prohibits VOA from spreading information in the United States. This was so that the U.S. government could not direct information at the American people, and so that VOA would not compete with commercial radio stations in the United States. The press can examine scripts and program materials at the VOA headquarters, but this information never leaves the building. This is why many Americans are not familiar with VOA. It is not publicized in the United States, and although books and articles are written about the station, it does not promote itself in this country.

On June 1, 1953, President Dwight D. Eisenhower signed the papers to create the USIA, the United States Information Agency. It took over the overseas information responsibilities from the State Department, including VOA. USIA was located in Washington, D.C., and it was felt that VOA should move from New York to the nation's capital for more effective management.

So VOA moved to Washington. The first program broadcast from their new studios was the Hindi Language Service program of September 23, 1954.

In March of 1959, a draft of a VOA Charter was written by staff members who felt that a written acknowledgment that VOA news was accurate and fair was needed. Although VOA began as a propaganda weapon, it was evolving into an information agency. By 1960, George V. Allen, the USIA Director at the time, approved the Charter, but despite this support and the support of the Presidents in the next sixteen years, it was July of 1976 before Congressional support was obtained and the Charter was finally made into law. Public Law 94-350 states: "The long-range interests of the United States are served by communicating directly with the people of the world by radio. To be effective, the Voice of America (the broadcasting service of the United States Information Agency) must win the attention and respect of listeners. These principles will therefore govern VOA broadcasts:

1. VOA will serve as a consistently reliable and authoritative source of news. VOA news will be accurate, objective, and comprehensive.
2. VOA will represent America, not any single segment of American society, and will therefore present a bal-

anced and comprehensive projection of significant American thought and institutions.
3. VOA will present the policies of the United States clearly and effectively and will also present responsible discussion and opinion on these policies.

The VOA Charter does not say what the radio programs must include, or how they should be written. It does give guidelines so that both VOA and the Congress understand what will be broadcast.

A Hindi language broadcaster at a VOA microphone

In May, 1985, VOA began a semi-independent organization broadcasting to Cuba and called Radio Marti. It was felt that more needed to be done to help promote the cause of freedom in Cuba.

In 1986, Senator Edward Zorinsky from Nebraska wanted to restate the Smith-Mundt Act prohibition against giving VOA information to the people of America. His amendment to the Foreign Relations Authorization Act did this, thus reinforcing the original intent. He pointed out that this law distinguishes the United States from the Soviet Union, where broadcasting propaganda to the people is a major government activity.

Today, VOA broadcasts news, music, and feature programs. Each language service develops its own programs except for the news and the VOA editorial. The editorial is written by the staff in the VOA Office of Policy and is broadcast daily on all language services. It explains the opinion and beliefs of the United States government. All services must use it, regardless of whether the broadcasters agree with it, or whether it is in agreement with the official policy of the country into which it is being broadcast.

VOA's aim is to provide information to the many peoples of the world who have no other way to get it. Letters from listeners show appreciation for VOA programs they hear. In 1985, VOA received almost 400,000 letters from listeners.

> "As a regular listener to VOA, nothing new escapes my ears. If you dial VOA, you take a ride on the eagle's wings and nothing escapes the sharp piercing eyes of the eagle."
>
> **LISTENER FROM GHANA, AFRICA**

"I want to thank you for all the activities of your broadcasting which have a very positive effect on a much larger number of listeners than you can imagine. It is good to know that on the other side of the ocean, there are still good and willing people who devote their work and time to informing us, unknown strangers, whose access to information sources is very limited. Because I get depressed by the negative influences of our surroundings, there are moments when I truly thirst for your cultural and enlightening programs. It feels like a remedy."

LISTENER FROM PRAGUE, CZECHOSLOVAKIA

★ ★ **TWO** ★

The Yankee Doodle Station

Although the Voice of America is a part of the United States government, it is a real radio station. VOA is different from commercial radio stations in some ways, but much the same in many ways. To get a better picture of VOA, let's compare it to a fictional American commercial radio station we will call WMCF.

• WMCF broadcasts only to the town in which it is located and the surrounding area. Every program is in the English language. WMCF is on the air every day from dawn to late evening hours.

★ The Voice of America broadcasts to the entire world, and many different languages must be used. The language services are on the air for varying amounts of time, from over 15 hours a day broadcast by the Russian Branch to the daily thirty-minute program of the Greek Language Service. Because of time differences around the world, some broadcasters must work late at night so that their VOA program airs during the day in the targeted country. Each language service works different hours.

• The call letters, WMCF, are mentioned often during the day to identify the commercial station. Musical jingles, using the letters WMCF, are played several times an hour.
★ VOA uses the tune "Yankee Doodle" in announcing its programs and in signing off. This makes VOA easily distinguished from other international radio stations such as the BBC (British Broadcasting Corporation) and Radio Moscow from the Soviet Union.

• WMCF can be heard by dialing 1213 on the AM dial of a radio.
★ VOA broadcasts mainly on shortwave radio. The different language services broadcast on different frequencies. Listeners can request a program guide from VOA headquarters in Washington, which gives information about frequencies and times of programs broadcast to various countries.

The Voice of America logo is red, white, and blue.

• WMCF uses commercials throughout the day. Businesses in the area pay for a 30- or 60-second commercial aired on WMCF. This money helps finance the station and its staff.

★ VOA is part of the government. It does not use commercials. VOA receives money from the U.S. government, and this funding must be approved annually by Congress. VOA staff members are government employees, and receive the same pay, vacations, and benefits that other government employees do.

• WMCF is known for its fast-talking disc jockeys and announcers who keep up a constant chatter between musical selections.

★ VOA uses music too, but announcers must talk clearly and distinctly. Most of their listeners have English as a second language. Broadcasters using foreign languages have to speak clearly also, taking care to pronounce words correctly.

• WMCF staff members—disc jockeys, engineers, newscasters, salespersons, the station manager—are all private citizens.

★ VOA has staff members who are announcers, engineers, newscasters, and disc jockeys, but it also employs Foreign Service officers and political appointees. Political appointees are men and women chosen for their job qualifications but also because they are in the same political party as the President and support his views. The Director of VOA is a political appointee, as is the Chief of Current Affairs and others.

• On weekends WMCF disc jockeys often make personal appearances at a new shopping mall or a store opening.

Equipment at VOA studios includes tape machines.

WMCF listeners enjoy meeting the radio announcers and discovering the real people behind the familiar voices on the radio.

★ VOA broadcasters have many fans and receive thousands of letters from listeners. But they cannot go to the local shopping mall and meet these people, since all their listeners are in other countries. However, when they visit countries where their programs are known, they are treated like celebrities. Willis Conover, a VOA announcer for over thirty years, is a hero when he goes to Poland. Yet he can walk down the street in America and not be noticed.

• To serve the listeners of WMCF, the station owners want to increase the time the station is on the air, hire another reporter for the news department, and sell more commercials.

★ VOA has begun a program to improve its services too. Some of its equipment and antennas are old, making it difficult for listeners to hear their programs. Over the next ten years, VOA will spend over a billion dollars in an attempt to make VOA programs louder and clearer.

One problem that WMCF never has is jamming of its frequencies. VOA broadcasts into some countries that do not agree with the views of the United States. In an effort to keep their people from hearing VOA programs, they broadcast chirps, bird calls, or just loud noise on the VOA frequency, knocking VOA off the air. Jamming has occurred since the beginning of Voice of America. At first, during World War II, it was done by the Germans and Japanese. In 1948, the Soviet Union jammed VOA for the first time. Countries currently jamming VOA broadcasts are the Soviet Union, Poland, Cuba, and Czechoslovakia.

VOA announcer Willis Conover is a celebrity in Poland.

The Soviet Union admits that it jams VOA and other radio stations. It defends the right to choose what its people are allowed to hear. Their argument is that governments have the right to control information coming to their country from other countries. VOA's position is that no government has the right to control information.

Jamming bothers many listeners, and they write to VOA to complain.

> "Western radio is the only way we can find out what's going on in the world. If it weren't for jamming, I would say VOA's programming is wonderful. Jamming is, however, always present nowadays, and must be taken into account. There was a time when young people recorded VOA's popular music broadcasts. I greatly enjoyed them myself. Nowadays listening to music is impossible, but VOA still broadcasts it. Listening to Western radio has become a conspiracy: it's better not to tell anyone you are listening, you never know what might happen."
>
> **LISTENER IN THE SOVIET UKRAINE**

> "I listen to Western radio to hear what the West thinks of world events and particularly of the USSR. The heavy jamming makes people lose patience and discourages them from listening."
>
> **LISTENER IN THE SOVIET UKRAINE**

So both WMCF and VOA are successful radio stations. Each has many listeners, good programming, professional staff persons. But VOA is an international radio station, and you sense that international flavor if you visit their offices in Washington, D.C. The Voice of America is housed in a building that looks like many other government buildings, but going up the escalator to the second

The Yankee Doodle Station ★ 19

floor you see the long, flowing sari worn by a woman broadcaster from the Bangla Language Service. You overhear snatches of the Arabic language from two men behind you. As you step off the escalator you are greeted by a display listing the word "Welcome" in over a dozen languages. In the hallway are framed photographs of the past seven U.S. Presidents speaking into VOA microphones. When you arrive at the master control room, there are two large world maps. A half dozen visitors from Japan soon arrive and join you in waiting for the beginning of the VOA tour. America is often called a "melting pot" of different people. VOA is a smaller version of this melting pot, blending the talents of people from many places to produce radio programs for listeners around the world.

Achala Sharma, a Hindi Language Service broadcaster, is seated in one of the many studios. At a signal, Sharma begins speaking into the microphone.

Sofa in front of control room where VOA tours begin.

The engineer at the controls monitors the sound as electrical impulses travel to the master control room, then to an antenna thirty miles away in Maryland. This antenna sends the impulses up to a satellite 24,000 miles above the earth. They bounce off and back down to a transmitter station in Kavala in northern Greece. As shortwave radio waves, they go up to the ionosphere where they are reflected and sent all the way to India.

People in India, including Achala Sharma's parents, can tune in to the Voice of America on their shortwave radios and listen to Sharma announce the VOA Hindi Language Program. It takes only a few seconds for the whole process, from Sharma speaking into a microphone in Washington, D.C., to people in India hearing her voice on the radio.

Achala Sharma broadcasts to India.

Music

When you listen to a radio station, what do you like to hear? If your answer is music, you are not alone. Many people in America and throughout the world enjoy listening to music. Some music fans switch stations when news or ads are on so that they can listen to nonstop music.

"Music is a universal language," says Judy Massa, the Music Director at Voice of America. In order to be a complete radio station and fill the needs of its listeners, VOA offers many music programs. Much of the music is American music. Listeners worldwide can enjoy the "Country Music USA" program, dance to the rock and pop tunes on "Music USA," and hear classical music on "Concert Hall." Many of the VOA language services have music programs that feature the favorite music of the people who live in the country they are broadcasting to.

In America, many disc jockeys become local celebrities

22 ★ The Voice of America

Judy Massa, Music Director of VOA, with Kenny Rogers

in the communities surrounding their radio station. VOA music announcers also enjoy great popularity and receive many fan letters. But their listeners are many, many miles away. The VOA music announcer cannot drive a video van into the neighborhood of his or her fans or appear at a local charity fundraiser as American announcers often do. Like many VOA broadcasters, some music announcers do not even give their real names on their VOA programs. This is because listening to VOA is against the law in some countries. If the announcers were to give their real names, it might endanger family members who still live in their homeland.

> "For a long time I have been meaning to write you, but the rough times prevented me from doing so. I do not know whether I should study or mourn the deaths of my friends. To dispel my sorrow over the brutalized youth who were forced to join the army, I have taken refuge in reading and in listening to music. I beg you to increase the airtime of Persian music and the 'Sound of Music' program to soothe the dejected hearts of the Iranian youth."
>
> **LISTENER FROM TEHERAN, IRAN**

Picture a gathering of seventeen-year-old teenagers in a house in the country of Iran, a nation once called Persia. Despite the fact that their government made music and clothing from Western countries, including America, against the law, the boys are dressed in American punk fashion. At 8:30 P.M. they tune in their shortwave radio to catch the beginning of their favorite program. An unidentified man says, "This is your announcer friend" as rock music fills the house and VOA's Farsi Language Service program, "Sound of Music," begins broadcasting to the people of Iran.

"Mr. Sound of Music" came to the United States in 1975 to go to college in Portland, Oregon. He studied communications, and made his radio debut at the college radio station. When the Iranian revolution turned his homeland into a land of war and torture in February of 1979, he stayed in America and continued his education. After earning his master's degree, he started working at Voice of America in the summer of 1981.

Since the revolution, Western music is banned in Iran and Persian popular music is strongly discouraged. This has been hard for the people of Iran, because before 1979, American and British rock and pop were widespread and popular.

The first Thursday of each month the "Sound of Music" becomes a call-in show. The VOA phone number is announced two weeks beforehand, and on that Thursday the announcer takes calls from listeners in Iran who request specific songs and records. The calls are recorded, and on a future date the taped voices and requested music are broadcast.

The Iranians call in on their home telephones or sometimes from public phones. It is dangerous because of the ban on Western music. Most callers do not identify themselves or they give fake names. It costs the Iranian youth $12 for a three-minute phone call to VOA in Washington, D.C., but the announcer says that the listeners are willing to pay the money because they want so badly to hear the music.

Sometimes, callers tell the announcer about current conditions in Iran. A caller from Kazerun, Iran, answered the question of how things were by responding, "Very bad. They are arresting kids everywhere."

Some devoted fans go to great lengths to call in their request. Hamzeh traveled over 400 miles from his home in Gonbad-i Qabus to the city of Teheran to make his phone call. Iran's international telecommunications center had been bombed, and Hamzeh wanted to be sure he would be able to reach VOA by phone. He figured he

Haideh, the best-known singer of popular Iranian music, is featured on Farsi Language Service programs.

would have a better chance by calling from a major city. He requested the latest song by Michael Jackson. There was excitement in his voice as he repeated several times, "Hamzeh's my real name. I'm not giving you a false name as some people do."

According to the announcer, who remains nameless because like all the Farsi announcers he does not mention his name on the air for safety reasons, his show provides a pleasant escape from reality for young people in Iran.

"The Iranians are very restricted, which only serves to increase their thirst and appetite for Western music. The listeners look forward to the entertainment on VOA, because they wouldn't otherwise get it. I get many letters from listeners, some with fake addresses. The people who write feel it is important enough to put themselves in danger; they are so fed up they don't care anymore."

The listeners who call in, mostly between fifteen and twenty-five years of age, usually request American and British rock songs they have previously heard played on "Sound of Music." Occasionally there is a request for Persian music. The VOA announcer is in touch with a group of Iranian exiles in Los Angeles, California, who play Persian music. He fills these requests with their music.

The "Sound of Music" program helps bring popular music into the country of Iran. It may not seem a form of protest for us to turn on a radio and dance and sing to some music. But in Iran, where protesting against the government can lead to severe punishment, young people listen to VOA in their private homes as a small sign of their independence. Try to imagine what it would be like to be denied the right to listen to music of your choice. In America there are dozens of radio stations which play music for every imaginable taste. Some people around the world do not have these choices.

Another music announcer on VOA helps the people of Africa discover their heritage and develop a pride in their native music. Every Sunday evening on the English to Africa VOA broadcast, hostess Rita Rochelle's melodious voice announces "It's Music Time in Africa." The strains of an orchestra from the African country of Sierra Leone are heard, and Leo Sarkisian's program featuring traditional and pop African music begins.

Sarkisian traveled an interesting route on his way to becoming VOA's "Music Man," as he is affectionately called by his many listeners.

He began as a commercial artist in New York, doing

Music ★ 27

Rita Rochelle and Leo Sarkisian, VOA's "Music Man"

illustrations for magazine covers and picture books. In his spare time, Sarkisian spent hours in the local library doing personal research on music. He had grown up with both Armenian and Turkish music, and he played traditional folk instruments. Sarkisian's interests included African music as well. Somehow copies of his papers got into the hands of a Hollywood producer, who visited him in New York and offered him a job as a sound engineer. In a matter of a few minutes, Sarkisian's career completely changed.

Since that day in the early 1950s, Sarkisian has made music his full-time job. Sarkisian went to Hollywood and worked as a sound engineer and music director for various films. He was then sent to the countries of Pakistan, Afghanistan, and Bangladesh to record music to be used in the backgrounds of movies. While in those countries, Sarkisian made a collection of all the traditional music he heard. It was there that he developed his special personal style that has served him well in his many VOA trips.

"I once visited a forbidden part of Bangladesh. Even the government had never been there. While I was still making arrangements, such as getting a canoe to take us up the river and getting my gear together, the people of the villages already knew I was coming. Because at some of the first villages we got to, they told me that they had a complete description of me. They knew that I was carrying equipment. They even described the earphones I carried. They said you put them on your head and you can hear our music. They knew this all ahead of time."

This did not make the people in the villages afraid of Sarkisian, however. He cannot recall ever getting a bad reception in any of the hundreds of villages he has visited over the years.

"They just accept you! They know I am not armed, that I'm not over there to look for other things, just for music. They get that feeling immediately because when I enter a village I spend a lot of time, sometimes two or three days, talking and listening to their music and getting up and dancing all night. I never take out my equipment right away. In this way, it helps me to know who the good musicians are, what they can do, and the sounds of their instruments. I get used to it, and I sort of wait and then they come asking me, 'Hey, when are you going to record us?' Then I take out my equipment. I learned this a long time ago. Some foreign recording companies take out their tape recorders right away, and say to the villagers, 'Sing' and all the musicians freeze up. You have to spend a lot of time with them."

Sarkisian then went to the country of Guinea, on the continent of Africa. While working there, he was offered a job with VOA by Edward R. Murrow, who was Director of the United States Information Agency at the time.

"Murrow told me that I wouldn't make money like I had in Hollywood, but I would have a ball. That's exactly what has happened."

"Music Time in Africa" first began in 1965 in Monrovia, Liberia, at the African Program Center there. It continues today with millions of listeners and hundreds of fan clubs, although now it originates in the studios in Washington, since there is no longer any broadcasting done outside of Washington, D.C.

> **"This short note is just to inform you that I am the best listener to your production 'Music Time in Africa' each Sunday on the VOA. My small radio vibrates with your golden voice and we enjoy the entire production because it concerns African traditions and culture. The selections of music are so good that all of us here in Cameroon love this distinguished program. God bless you for not forgetting us here in Africa."**
>
> **LISTENER IN BAMENDA, CAMEROON**

Sarkisian, who has visited every African country except Zimbabwe in his efforts to preserve Africa's musical heritage, has inspired unbelievable fan loyalty.

When he traveled with Rita Rochelle, the popular hostess of the program, to Africa in 1985 to join in the celebration of the Ghana Broadcasting Corporation's 50th anniversary jubilee, they made a stop in Liberia. Only an hour after they checked into their hotel, they were told they had two visitors. In walked two fans, very dusty from having walked nonstop for five days. They had heard of Sarkisian and Rochelle's impending visit on one of the VOA broadcasts, so one of them walked from his home in the southern part of Guinea across the border into Liberia, where he picked up his friend and continued the long journey to meet the "Music Man."

On his program, Sarkisian talks about traditional African music and instruments. When planning each show, he takes care to include music from all parts of Africa. This is important to the listeners. Although there are many radio stations in each African country, VOA is the only station that gives one geographic area of Africa a taste of the other areas.

African music is different from American music. African music is made up of singing, a drum rhythm, hand clapping, and dancing. If any of these elements are missing, Sarkisian says, it is then not African music. His VOA program supplies the first three ingredients, and the listeners supply the dancing when they gather in groups Sunday nights around the shortwave radio. Of course, since Sarkisian's tapes have been recorded in the villages themselves, listeners occasionally hear other unusual sounds such as a baby nursing as her mother sings into the microphone.

"Music Time in Africa" is broadcast to Africa, but can be heard around the world as part of the VOA English Service. It is also translated into Portuguese and French and broadcast to parts of Africa where those languages are spoken.

Sarkisian has brought pleasure to many people during

his years with VOA. And he has learned something important as well.

"When you go to the industrialized countries—London, New York City, Paris, Tokyo—all the streets are paved and all are concrete. The people move very fast. Sometimes it's very difficult to feel the earth or touch the earth. But when I go to a village and the village elder and I sit down and begin to speak, while he is talking I can touch the earth. That's reality, and that's what Africa has taught me."

Jazz is another type of music featured on VOA. Many people, including both musicians and fans, disagree as to what jazz is, and what songs can be considered jazz music. But jazz buffs the world over agree on the identity of one of the greatest friends of jazz. That person is Willis Conover, the host of "Music USA" on VOA.

Since January of 1955, VOA listeners worldwide have enjoyed his deep, smooth voice introducing the best of jazz and American standards. To many people, especially in the communist countries of the Soviet Union and Poland, the beginning of his theme music, Duke Ellington's "Take the A Train," is a signal to stop whatever they are doing and listen. Conover has been bringing his program, "Music USA," to music lovers for over thirty years. His daily workload would exhaust many other people, but Conover has a belief in the music he plays that goes beyond the words and tunes. To him, jazz parallels American life, and is a freedom that is available to everyone despite the laws and government they may live under.

"Music USA (Jazz)" is a 45-minute program aired daily Mondays through Saturdays. On Sundays and Mondays, Conover also hosts the 30-minute program "Music USA (Standards)." Because of the preparation time involved, this means many long hours in the VOA studios.

Each program is carefully planned. The songs are chosen to fit with the song before and the one following. Conover builds upon a theme, and uses the music to convey feelings of happiness, sadness, and other emotions. His programs are not like some American rock music shows where the disc jockey is very upbeat and chatty, filling every available second with fast talk. Conover talks in slow, sustained tones, introducing the music and then letting it speak for itself. When you hear Willis Conover on VOA, you get the feeling that he is in your living room discussing his favorite songs, not in a radio studio in Washington, D.C., speaking into a microphone.

> "This is just my way of saying how much I appreciate your program, 'Music USA Jazz and Standards.' I have been a regular listener to the show for the past twenty years. I find it very informative as far as the performers are concerned, and also very entertaining. I think your government is doing a splendid job in trying to bring the world's people closer together through music."
>
> **LISTENER IN TRINIDAD, WEST INDIES**

Conover is no stranger to commercial radio. The son of an Army officer, Conover went to many different schools as a child. He played the part of a radio announcer in a class play at one of the schools. Enjoying that experience, he got a job at age nineteen in a radio station. He began to develop his love for jazz music then, occasionally slip-

Music ★ 33

Willis Conover, host of "Music USA (Jazz)" with Duke Ellington

ping in a song by Louis Armstrong or Duke Ellington.

Conover has traveled and met many of his listeners. In the spring of 1959, he first visited Warsaw, Poland. As he looked out the plane window shortly before landing, he noticed a 40-piece band and many people waiting at the airport gates. To his surprise, the big reception was for him! There were two concerts during his visit. The musicians had gathered at their own expense to play for their friend, Willis Conover, to show him what they had learned from listening to his program.

When Conover made his eighteenth visit to Warsaw in 1984, he was again met at the airport by a band and television reporters. Conover is immensely popular around the world. One listener from the Soviet Union wrote that Conover was a source of strength when he was overwhelmed by pessimism. Another listener wrote, "The world changes, leaders die, governments fall, but every night you turn on the radio and there's Willis!"

But perhaps this story best illustrates how a listener's letter can change a life.

Willis Conover was interviewed for an article that appeared in the *Washington Post* in November, 1984. He was sent a letter from a woman who had been astounded to read about him in the *Post* article. Fourteen years before, Evelyn D. Tan had lived in Manila, Philippines, with her family. She was sent to live with her cousins in China that year, but there were many political problems and it was a dangerous place to be. In an effort to forget about the problems, she listened to Willis Conover's jazz program on VOA and was introduced to Duke Ellington and other great jazz musicians. The very act of listening to VOA was daring for a young woman living in China in the early

1970s, especially since the security police were just two blocks away from where she lived.

When Tan's letter was received at VOA, Conover's assistant wrote her back. She thanked Tan for the letter and explained that Conover was ill and in the hospital. Tan then sent Conover a get-well card. Thinking that he could not just ignore such a considerate fan, Conover called Tan when he got out of the hospital. He thanked her for the letter and card, and invited her to VOA headquarters to see his studios. She visited VOA on November 20, 1984. They were married on May 23, 1986.

For over three decades, Willis Conover has passed on his love of jazz music to many friends around the world. To some listeners, "Music USA" is merely entertainment; to others, it means much more.

"I'll stay your friend forever," wrote a listener from Moscow, Soviet Union, "because you are the man who opened the window to the jazz world for me and made me happy for the rest of my life."

★ ★ **FOUR**

News and Features

The telephone rang. Bill Royce, the Chief of the VOA Farsi Language Service which broadcasts to Iran, picked it up and began a spirited conversation in Farsi. He was speaking with someone from one of the Iranian opposition groups in Paris, France. The caller told Royce that there was a doctors' strike in Iran. He reported that 99 percent of the doctors in that country didn't go to work that day.

Royce had a pretty good feeling that this story was accurate. He called a Farsi Language Service news staffer into his office. The staffer wrote a short report about the doctors' strike, and took it to the VOA newsroom. This was an important story; the VOA listeners in Iran would be very interested in the strike and would benefit by hearing details on its impact across the country.

The newsroom could not find this information about the doctors' strike anyplace else. The news services had not yet reported it, and other government agencies knew nothing about it.

Did the VOA Farsi Service get a scoop and report this information before anyone else? No, they didn't; in fact, they waited until they got another reliable news station to confirm the information before they even mentioned it on the air.

Have you ever heard some gossip or a shocking piece of news from a friend? You may have repeated this interesting news to another friend, without checking whether it was true. If it turned out to be false, you were probably very embarrassed for having passed on incorrect information. You may also have hurt someone's feelings and lost friends in the process.

This explains the VOA second source rule. News items have to be reported by at least two sources (these are news organizations, radio stations, or people like the man in Paris who are honest and directly involved in the situation) before VOA will broadcast them on their newscasts. This may mean that VOA is not the first with the news, but it also means that the news on VOA is true and accurate.

> **"I am a frequent listener of your Spanish broadcasts and congratulate you on their quality. Your news greatly interests me and is the only news that offers truthfulness."**
> **LISTENER IN DOMINICAN REPUBLIC**

When the Farsi Language Service finally reported the doctors' strike, VOA knew it was a true story. The listeners in Iran were able to get timely information that they could trust. As one listener said, "If VOA says it, we know it is true!"

News is very important at VOA. Most listeners overseas tune in to hear the news. They expect and appreciate accurate, timely news reports.

Let's follow a news story from when it happens to when it is broadcast on a VOA news report.

• VOA White House news correspondent Philomena Jurey attends a press conference at the White House and hears President Reagan make an important announcement about children and drugs. He discusses the growing problem of drug abuse, quotes figures about how many children use drugs, and urges youngsters to say no to drugs. One of his goals in his plan to stop drug abuse is international cooperation, which would certainly affect VOA audiences.

• After his announcement, President Reagan answered questions from the assembled reporters. Jurey does not raise her hand to question President Reagan; she is busy taking notes. A USIA policy prohibits VOA news correspondents from asking questions during presidential news conferences. This is so that the VOA correspondents do not compete with the other correspondents for the limited question time. Jurey returns to the booth she shares with a VOA engineer in the basement of the White House, and types a summary of the press conference and the important announcement into a computer.

• She runs off this summary onto a piece of paper (this is known as hard copy), and then the engineer calls the central VOA news studio and says she has a report to transmit. (This studio is called the "bubble," because it has glass all around it.) The engineer operates the equipment to broadcast her report.

News and Features ★ 39

Pictures of Washington-based correspondents at VOA headquarters

• When the person in the "bubble" says, "Go from the top," Jurey reads her report. She first says the name of the report, then she does a countdown (five-four-three-two-one). One second after Jurey says "one," she begins her report. The report is only a minute long. At the end Jurey says, "This is Philomena Jurey, VOA News, the White House."

• The engineers in the "bubble" record this report. It will be used on English broadcasts. When it is used, it will

VOA studio known as the "bubble"

be introduced: "Reagan's announcement on children and drugs is the subject of this report filed by VOA correspondent Philomena Jurey at the White House."

• Jurey then has to file the written report. The languages other than English cannot use Jurey's voice report. They need a written report that they can translate into their language. The computer that Jurey used in the booth at the White House is electronically connected to the newsroom at VOA. So when a certain number is dialed, the words from Jurey's report come flying into the computer in the newsroom. It is checked for typing errors, and then a hard copy is run off.

• If for some reason Jurey couldn't send the report through the computer, she would have read the report to a person in the newsroom who would have taken it down. Some words that might cause problems when translated into another language are spelled out. This is especially important for the names of people and places. (Jurey rarely

Philomena Jurey writing a report on location

needs to spell out words, but news correspondents stationed overseas have many occasions to do this.) The International Phonetic Alphabet is used to spell out these words. For the word "Reagan," Jurey would say, "There is one spelling, the name Reagan. That's R as in Romeo, E as in Echo, A as in Alpha, G as in Golf, A as in Alpha, and N as in November."

The International Phonetic Alphabet is:

ALPHA	NOVEMBER
BRAVO	OSCAR
CHARLIE	PAPA
DELTA	QUEBEC
ECHO	ROMEO
FOXTROT	SIERRA
GOLF	TANGO
HOTEL	UNIFORM
INDIA	VICTOR
JULIETTE	WHISKEY
KILO	X-RAY
LIMA	YANKEE
MIKE	ZULU

At the same time Jurey is filing her story, reporters from other news services, such as the Associated Press (AP) and United Press International (UPI), are filing their own stories on President Reagan's announcement. These stories arrive at VOA on machines that print reports on long rolls of paper. The VOA editors compare these reports to the report from VOA correspondent Jurey. Sometimes one of the news service reports may mention additional details. Other radio news reports are also checked, and all the major television stations are monitored.

News and Features ★ 43

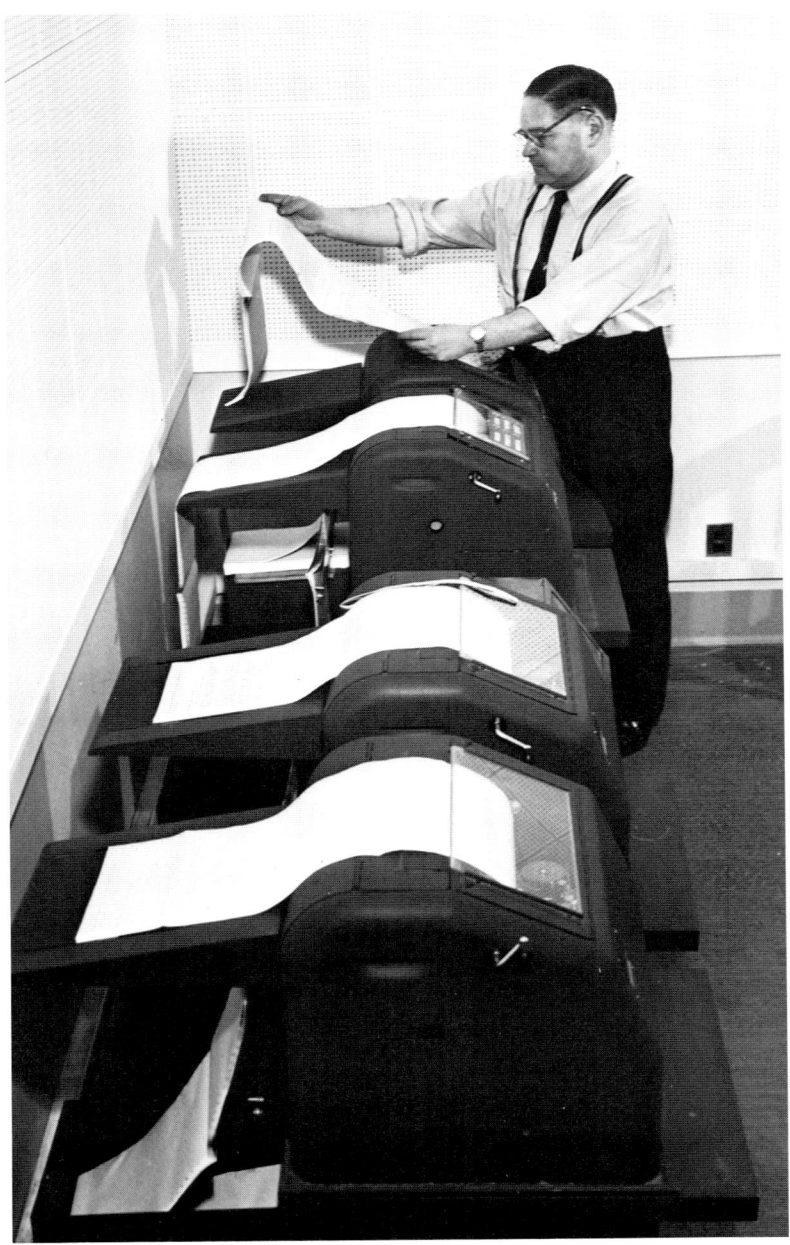

News comes into VOA studios on teletype machines.

All these versions of the same story are given to a regional desk in the newsroom. Each regional desk covers a certain area of the world. For example, there is a desk for Latin America, and another one for South Asia. There is at least one writer and an editor at each regional desk.

The writer decides to write a one-minute story on the President's announcement that will be 12 typewritten lines long. She uses information from all the different reports in writing her story.

This written story then goes to the regional desk editor, who reads it. Perhaps he doesn't like the "lead." The lead is the first sentence of the story containing the basic who, what, when, where, why and how information. The writer would then have to rewrite the beginning of the story. The editor also makes sure that the story is just reporting the facts, and not taking a position on children and drugs. VOA writers only report on what other people say and events that happen. They are careful not to include any personal feelings about the subject.

The regional editor then sends the story to a central "duty" desk. There are two more editors there, a copy editor and a duty editor. The copy editor is the next to last person to check the story for errors; the duty editor is the last person to make sure the story is correct and complete. If the Director of VOA comes into the newsroom and questions why it took so long to get the report about the President's announcement on the air, or wonders how the number of children who use drugs in America was decided, he will talk with the duty editor.

The story is then sent up to the language service that is going to use it, where it is translated into that language and read by the broadcaster on the next newscast.

News and Features ★ 45

In addition to Philomena Jurey, there are other full-time VOA news correspondents in the United States, and also VOA correspondents in other countries.

All news stories come from the central VOA newsroom. If a Spanish Language broadcaster hears something at lunch, he must bring that information to the newsroom to be checked out and confirmed before it can be broadcast as news. Every two hours throughout the entire day, the newsroom sends the language services a list of twelve to fifteen top news stories that may be used in their newscasts.

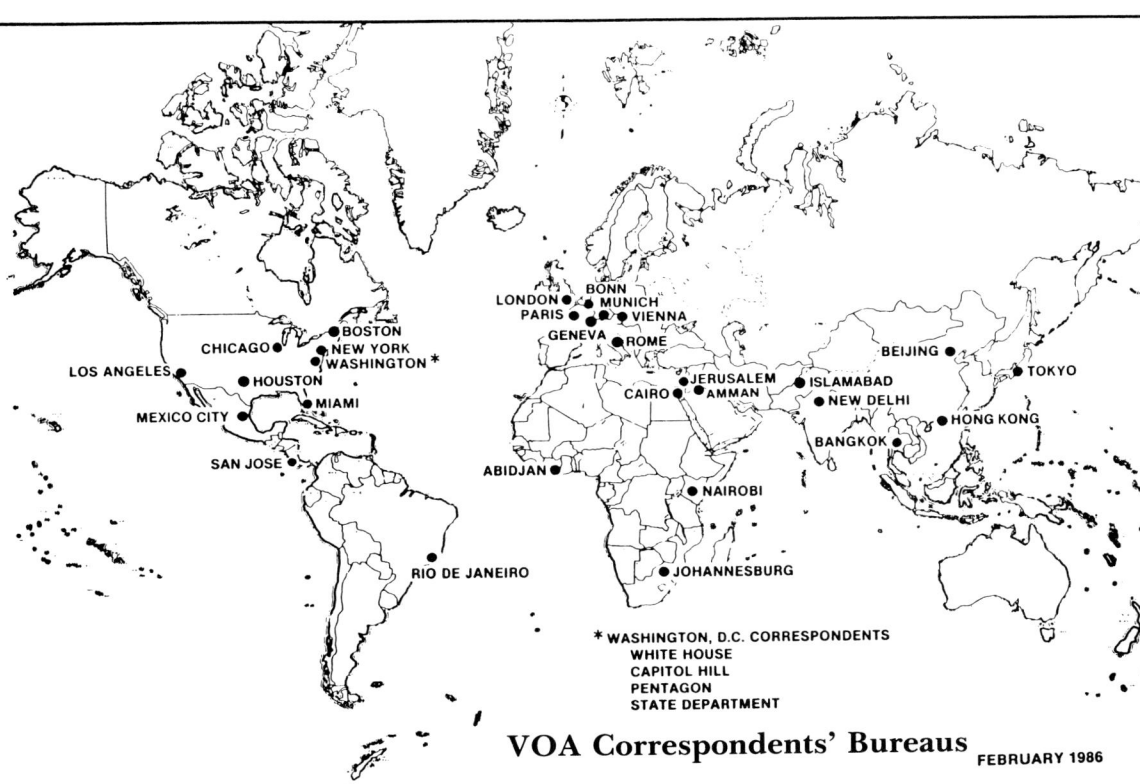

Voice of America correspondents are stationed around the world.

Working as a radio journalist can be fun. Philomena Jurey is a member of the White House press corps, and she follows the President every day, even when he is away from Washington and traveling around the country or world. When Jurey accompanied President Reagan on his visit to China several years ago, she was a celebrity. The Chinese did not know the correspondents from the network television stations or the other reporters, but they recognized Jurey's voice and asked to meet her. When Philomena Jurey's reports are broadcast over the English Language Service, the report she sent to the "bubble" is used, so her voice is heard. Many people in China listen to the English Language programs.

As a professional journalist broadcasting to people in other countries, Jurey thinks about the impact some of her stories have and whether they are putting America and the presidency in a bad light. But she still reports the truth, no matter how it makes Americans seem to other people.

"You can't go wrong if you just tell it like it is, it's the only way to do it."

When President Reagan left on a trip to Tokyo and Indonesia, he was asked if he was worried about any terrorist acts against him or Mrs. Reagan. He answered that he wouldn't even talk about it because he was superstitious. Jurey mentioned this comment in a story that was broadcast over the English Language newscasts. She was asked by a VOA intern whether she was right in quoting the president saying he was superstitious, because people in other countries might think that is a sign of weakness or failure.

"No," Jurey replied. "My duty is not to make a judgment

about whether it was the right thing or the wrong thing for him to do, my duty is to report what he said."

Another way to collect stories and features about America is to actually travel around the country and report on the people you meet and the places you visit.

The VOA Voyager van does this. It is a red, white, and blue van that has a fully equipped broadcast studio in it. The Voyager has been traveling across the United States recording programs about America since January 18, 1985. It has a microphone, tapes, and everything else a regular studio has. Programs can be produced in the Voyager and sent to Washington to be used immediately, or taped for future use.

The VOA staff in Washington decides where the Voy-

The VOA Voyager van at Mount Rushmore

ager will go each week. This schedule is given to the language services, and language service broadcasters and English service reporters travel with the van each week. An engineer and staff person from the VOA Special Events Division also accompany them. They only travel in the van for one week at a time because they do not want to be away from their other work any longer than that. They choose which week to go according to the needs of their particular service. For example, the Dari Language Service, which broadcasts to Afghanistan, might send a broadcaster on the Voyager when it visits Seattle, Washington, because there is a large Afghan community there.

Several weeks before the Voyager visits a state, the VOA staff makes contact with local tourism bureaus, chambers of commerce, and the state capital to find places that the van should visit. Sometimes it ends up in a town the VOA staff has never heard of before.

This happened in Fergis Falls, Minnesota. After talking with people who lived there, the staff was able to plan a very interesting trip. The van visited the only person in the world who makes a unit for clipping the nails of horses—gigantic nail clippers. They also found a farm where llamas are raised. So the Voyager had a wonderful schedule that day, and the broadcasters on the trip learned a lot.

These stories give VOA listeners glimpses into the lives of ordinary Americans. It is hard to do stories like this if you spend every day in Washington, D.C. So the Voyager van goes from coast to coast to put American people on the air, so that the rest of the world can hear what they are like. A listener from Beijing, China, wrote, "VOA is a window through which I can look around a strange continent."

VOA broadcasters in front of the Voyager van

Also, traveling on the van and meeting Americans is an educational experience for the broadcasters, most of whom originally lived overseas. They can discover that America has mountains like the mountains in their homeland. They can meet Americans of every color, educational background, and religion. The broadcasters and listeners learn that there are many Americans, and that they do not speak with one voice. Americans in North Dakota think differently on issues than people in Florida or Virginia.

Another benefit from the Voyager van is the exposure about VOA it gives to Americans. When VOA was six years old, Congress passed a law that said VOA could not give information to the people of the United States. This prevented an administration from using the VOA for its own political purposes by broadcasting their ideas to the people in America.

But when the Voyager van visits an American city, it becomes a local news story for the reporters of that city. The VOA broadcasters are interviewed, and asked questions about VOA, and about themselves and their lives.

When the Voyager visited Temple, Texas, on July 31, 1985, it was declared Voice of America Day and the VOA correspondent from the Czechoslovakia Language Service traveling with the van was named Honorary Mayor. There was a press conference at city hall attended by the newspapers and the two local television stations.

Special English

Imagine you are studying a foreign language in school. You have been doing well in the class. You can read all the sentences in your French book, and you understand everything on the language tapes in the lab. Deciding you are ready to move beyond the classroom, you go to a fancy French restaurant for dinner. There you discover how little you really know! The waiters are speaking French so fast you are unable to understand a single word.

Many people around the world study the English language. But as you discovered, knowledge of the grammar and vocabulary of another language does not guarantee that you can understand a conversation spoken in the normal, fast pace.

This explains the appeal of the Special English programs on VOA. Many people around the world want to listen to radio programs in English. English is not their native language, and although they are studying it they are not proficient and cannot understand regular English radio programming.

Special English programs are broadcast for people who are just learning English. Special English is not a new language, it is simple American English, designed for people who are not fluent in the language.

Only about 1,500 different words are used, and the sentences are short and direct. Special English programs are read at a much slower pace than other English broadcasting. The Special English broadcasters speak only 90 words a minute, which is much slower than the 130 words a minute most of us speak.

The Special English division at VOA takes regular news items, stories, and other written materials and translates them into Special English. The writing is not simplified to a child's level, but the writers sometimes reword sentences so that only short, basic words are used to convey the same meaning.

Next time you hear an adult conversation, pay attention to the vocabulary used. Adults may know long, fancy words but they seldom use them in day-to-day conversation. Most American adults use the same 1,500 words that are used in Special English programs.

Four hundred million people claim English as their second language, and millions of people are studying English in many countries around the world. Special English programming increases their knowledge of the English language. However, it is not intended as an English language lesson. Many of the regular language services of VOA offer English lessons already. The purpose of Special English is to communicate news and information to listeners who are trying to learn English and increase their English conversational skills.

Special English ★ 53

> "I'm a college student in the English Department. My English was very poor at first, then I listened to the Special English of VOA every morning and evening. Now I can understand 70 percent of it. Thank you, Special English! It has helped me with my English. I'll continue to listen to it. Many of my classmates listen to it, too."
>
> **LISTENER FROM ANHUI PROVINCE, PEOPLE'S REPUBLIC OF CHINA**

Special English programming was started on VOA in the late 1950s. In studying how to reach listeners to whom English is a second language, VOA put together a group of words that are essential to the English language. The first Special English programs, originally broadcast only to Europe and the Middle East, went on the air on October 1, 1959. The reaction of the listeners was so enthusiastic that Special English increased its area of coverage so that now almost everyone in the world can hear it.

> "I am living in a remote part of my country. In my workplace there are no proper facilities like light or piped-in water. The newspaper does not even carry important international news. I walk a distance of 2.5 kilometers to listen to VOA news in Special English. With your English to Africa program, I can now follow the events closely."
>
> **LISTENER FROM CROSS RIVER STATE, NIGERIA**

Announcers have special problems broadcasting Special English programs. Like all announcers, they need to speak very slowly. The announcers don't want to sound boring, though, so they have to remember to keep expression in their voices despite the slow pace. Writers also have to make some adjustments. Because the reports and features are read so slowly, you cannot use as many words in a Special English piece as you would with regular English. A four-minute report has to have a very short script, yet the important facts must still be included.

Special English news is broadcast on VOA on the half hour. Four-minute science features are on Mondays through Fridays, and five-minute features on words are broadcast on Saturdays and Sundays. In addition, a special feature program is broadcast each day.

In response to many requests from listeners, the Special English division prepared the *Word Book*, a dictionary of the 1,500 words used in the broadcasts. In addition to the pronunciation and meaning of these words, the book also contains tips on how to listen to a foreign broadcast and sections listing the months of the year, parts of the body, numbers, science facts, and information about the government of the United States.

VOA also has a popular book, *Words and Their Stories*. This is a compilation of scripts from the weekend features on words. It is a fascinating glimpse of the origins and meanings of some of the most colorful elements of American English. Listeners can write in to VOA and request these publications.

Would you like an explanation of the word "gobbledegook," or the phrases "to talk turkey," "chip on the shoulder," and "to break the ice"? Do you know the mean-

Special English ★ 55

Special English display at VOA headquarters

ing of "shilly shally" and "to talk through your hat"? Without an explanation, these phrases and words could have entirely different meanings to someone from another country who translates everything literally. The Special English division has received letters that have used some of the phrases from *Words and Their Stories,* such as the letter thanking VOA for "getting down to brass tacks."

The staff of eleven in the Special English division receives nearly three hundred letters from listeners every month. Many people have credited the Special English broadcasts for helping improve their English language skills. Almost half of the listeners live in China, and many teachers there record the Special English programs to use later in their classrooms. The Chinese published an edition of the VOA *Word Book,* which sells for 15¢ in their money and eliminates the sometimes long wait to receive the book from VOA headquarters in Washington, D.C.

> "I am Chinese. I came to Lulea to study for a Doctor's degree on October 6, 1985. First, I would like to express my thanks to you, VOA, with all my heart, which I can't describe here in a few words. I began to improve my English by listening to your voice, your unique 'Special English' program in July, 1982, when I was in the second year at the University. You can see what progress I have made since then because now I am here in Sweden studying for a doctorate. And the language we use in communication is English. According to Chinese traditions, I should say that this success should be attributed to you to a great extent."
>
> **LISTENER IN LULEA, SWEDEN**

It is hard to appreciate Special English without hearing it spoken, because the slow delivery is part of its very essence. The excerpt below is first given in standard English as it appears in a book entitled *Norse Stories* by Hamilton Wright Mabie. Then it is given in Special English. Note the changes made to adapt it into Special English.

Try reading the Special English version aloud. Since it must be read at the rate of 90 words per minute, this passage should take you about 100 seconds. Set a timer or have someone time you while you read the passage. Remember to talk slowly and clearly with lots of expression, and you will discover what Special English is all about.

ORIGINAL VERSION

"It was evening when the bride came driving into the giant's court in her blazing chariot. The feast was already spread against her coming, and with her veil modestly covering her face she was seated at the great table, Thrym fairly beside himself with delight. It wasn't every giant who could marry a goddess!

"If the bridal journey had been so strange that any one but a foolish giant would have hesitated to marry a wife who came in such a turmoil of fire and storm, her conduct at the table ought certainly to have put Thrym on his guard; for never had a bride such an appetite before. The great tables groaned under the load of good things, but they were quickly relieved of their burden by the voracious bride. She ate a whole ox before the astonished giant had fairly begun to enjoy his meal. Then she devoured eight large salmon, one after the other; and having eaten up the part of the feast specially prepared for the hungry men, she turned upon the delicacies which had been made for the women, and especially for her own fastidious appetite."

SPECIAL ENGLISH VERSION

"The goddess arrived at the giant's house late in the day. The giant, Thrym, had a huge feast ready to honor her. He was excited and happy because she had come to marry him.

"Her arrival in a most unusual vehicle, a chariot of fire, did not trouble Thrym. After all, how many giants got to marry a goddess? What happened when the goddess began to eat, however, should have worried any future husband, even a foolish giant. Never had anyone eaten so much, so quickly. The goddess ate a whole cow before Thrym had more than a few bites of food. Then, as the giant and the others watched in surprise, she ate all the fish, eight large salmon, one after the other. She left nothing for all the hungry men to eat. Then, she began to eat the food that had been specially prepared for her and the women."

★ ★ **SIX** ★ ★ ★ ★ ★ ★ ★ ★ ★

Children's Programming

The Voice of America does not do very much programming for children. There are reasons for this. Most people tune in VOA to hear the news. Knowing this, VOA does a lot of newscasts and features about the news. Since children do not usually tune in radio stations to hear the news, this means that VOA is not programming for the younger listeners.

Another reason concerns the safety of listeners. Since listening to VOA is officially discouraged in many countries, some VOA language services don't broadcast children's programming because they do not want to put the younger listeners in danger.

Some language services do not program children's features because this type of programming is already available on other radio stations in the country they are broadcasting into. VOA does not want to duplicate programs that are already on the air. VOA tries to give information that is unavailable elsewhere, especially information about America.

But some language services do have programs especially for children. One of these is the Dari Language Service which broadcasts to the people of Afghanistan, an Asian country bordering the Soviet Union, Pakistan, and Iran.

There is a program of children's stories. The announcer is particularly qualified to host this program. Mrs. H, who does not use her real name on the air for safety reasons, taught children for many years in Afghanistan, and she has seven children of her own. Her life story is as interesting as the stories she reads on the air.

Mrs. H was only three years old when her family left their homeland of Afghanistan because of a civil war. After twenty years in Iran, she returned with her mother, brothers and sisters. She became a teacher, and worked with children in the first through tenth grades. At the same time, Mrs. H became the first woman on Radio Afghanistan. All the women in Afghanistan wore veils across their faces. Mrs. H did also, but she had to move the veil aside when she spoke into the microphone at the radio station so that her voice would not sound muffled. Many years later she received a letter from the Foreign Ministry giving her permission to remove her veil for broadcasting.

Mrs. H read the news on the radio in Afghanistan for thirty years. When the Soviet Union invaded her country, she felt uncomfortable broadcasting the news. So she told the radio station that she was ill and could not work until she got better. After one year, she was visited at home and asked to resume her broadcasting duties. It was early morning, and she had just woken up.

John Denver, top, and Charlton Heston have appeared on VOA. Celebrities like these are enjoyed by both youngsters and adult listeners.

"I will come," Mrs. H whispered hoarsely as she rubbed her eyes. She pretended to have problems with her throat and eyes because she still did not want to broadcast in her country with the communists living there.

At that point, Mrs. H and her husband made the big decision to leave their country. She had secretly listened to VOA, and she wondered if she could possibly get a job as a VOA broadcaster. But the first step was to leave Afghanistan and go to Pakistan. They began their journey in their own car, and then used an abandoned Russian jeep. This made it easier to get through the tanks and patrols they passed, but it almost cost them their lives.

"We were captured by the resistance, and had to show our marriage papers to prove who we were. After 48 hours they let us go. They wanted to kill us if we were Russian because the Russians had destroyed their village."

In Pakistan Mrs. H and her husband had their freedom, but life was hard. They were there for ten months before being accepted as refugees to the United States. They ran out of money, and finally Mrs. H sold her wedding ring in order to survive. In February of 1982 Mr. and Mrs. H arrived in Seattle, Washington. It was a new country, a new home, yet Mrs. H. was reminded of her homeland. "The mountains in Seattle looked like mountains in Afghanistan."

Mrs. H became a VOA broadcaster, and because of her extensive radio experience she was well prepared for her new American job.

People in Afghanistan remember Mrs. H. She does not use her real name on the air, in order to protect loved ones in her homeland. Like other broadcasters from countries where listening to VOA is forbidden, she does not

Children's Programming ★ 63

VOA cares about children. In 1975, a day care center was started at Tinang in the Philippines where the VOA relay station is located.

want to provoke the authorities. But she receives many letters from people who recognize her voice and are happy to hear her on the radio once more. Now, with the children's program on VOA, Mrs. H is broadcasting to a new generation in Afghanistan.

"Flowers of Life" is a children's program broadcast by the Farsi Language Service to the young people in Iran. Mrs. M did children's programming in Iran. She was on the radio and television stations in Iran daily for twenty-five years.

"Flowers of Life" has three parts. The first part is news of children's events in the United States. Mrs. M finds information for this from newspapers and correspondents' reports. The second part is American music, and the third part of the program is storytelling.

Mrs. M reads stories that she knows the children in Iran will enjoy. She sometimes receives letters from young listeners requesting certain stories. One listener asked for the story of E.T., the Extra-Terrestrial of movie fame. Stories about space are always popular.

Mrs. M reaches out to Iranian children in the United States too. Recently the Textile Museum in Washington, D.C., had a special exhibit of rugs. These rugs had pictures on them that illustrated stories of the history of Iran. At Mrs. M's suggestion, the museum set up a program for the children in the Iranian community of Washington, and Mrs. M told them stories in the Farsi language.

Some of the regular VOA programs attract younger audiences, mostly teenagers. They enjoy listening to the music programs, and some younger people listen to VOA to improve their English language skills.

Sometimes, programs explaining the history of a country indirectly help children. In some countries occupied by the Soviet Union, knowledge of the country's history is scarce. Parents do not know about their own homeland's heritage. History programs broadcast on VOA help these parents teach their children about their homeland.

"I really do not know whether you can imagine the difficult situation faced by my parents in Czechoslovakia. They must be very careful what they tell their children so that it is not different from what they learn at school. That is the situation up to the tenth year of their children's age. . . . When the comrade teacher quits being the highest authority for the child, then the parent can start clarifying and explaining a few things. However, a good majority of today's parents were already born in the 'people's democracy' and they themselves do not know much about the history of prewar Czechoslovakia, and they themselves need to learn. You, the VOA, are providing this missing gap in their education by airing programs about the history of modern Czechoslovakia."

LISTENER IN CZECHOSLOVAKIA

★ ★ **SEVEN** ★ ★ ★ ★ ★ ★ ★

Focus on the Soviet Union

The Voice of America has a Soviet Union Division, which broadcasts in six languages—Armenian, Azerbaijani, Georgian, Russian, Ukrainian, and Uzbek. Every day these language services broadcast news and feature programs to the people of the Soviet Union. The people there sometimes cannot get news from any other source. Often the news they do get from the newspapers and radio and television stations in their homeland is different from the news broadcast by VOA.

> "You can learn a lot from our press and radio, but not everything. Western radio broadcasts additional information. I don't know if what VOA says is true, and I can't check. What VOA says about Afghanistan, for example, contradicts Soviet media reports. It's difficult to know who is telling the truth."
>
> LISTENER IN THE SOVIET UNION

> "Western radio is virtually indispensable. Soviet radio only serves the needs of Soviet propaganda. Both the national and the international news are so distorted that we don't even know what's going on in our own country."
>
> **LISTENER IN LATVIA**

People in the Soviet Union live under a different form of government from ours. They are curious about the differences between their country and the United States. Some of the most popular programming there has been VOA programs that tell about the lives of everyday Americans. "Work and Daily Life" gives information about jobs

All VOA programs, including those going to the Soviet Union, go through the master control room.

held by Americans, with details about pay, benefits, and pensions. In a program about American farmers, mention was made of a farmer's son being away at college. The fact that a farmer could afford to send his son to college impressed many listeners in the Soviet Union. This proved to them that an American farmer was living well. If it had merely been stated that American farmers live well, the listeners probably would not have believed it.

> "Any normal person wants to know what is going on in the world; how other people live, how they spend their free time, and how they work. I like VOA and find no faults with it."
> **LISTENER IN THE SOVIET UNION**

> "I am a cultured person, not a sheep. If you listen to Western radio, you learn new information, and you respect yourself for it. VOA is jammed, but those who really want to listen put up with it."
> **LISTENER IN GEORGIA, SOVIET UNION**

There is no VOA news correspondent in the Soviet Union. All of the news broadcast by the Soviet Union Division is obtained from wire service reports from within the country or from reports from VOA correspondents in neighboring countries. Many feature programs are aired. Some of these include Russian writers and artists who are now living in America.

Focus on the Soviet Union ★ 69

Alexander Toradze is a Russian pianist who has appeared on VOA.

The news and features about America tell the truth. Sometimes this makes America seem less than perfect. But the broadcasters feel that all sides of our country must be shown to the people in the Soviet Union. "Only by mentioning the warts can we make people believe that overall our complexion is really lovely," said one broadcaster.

The Russian Language Service broadcasts a youth show, aimed at children thirteen and older. This program features music and discussions of topics that interest that young audience, such as clothing styles. It is estimated that there are about 30 million regular listeners of VOA in the Soviet Union. Many of them have gone to great lengths to listen because the jamming is so strong. VOA seems to fill their need for news about their own country and the rest of the world.

President Ronald Reagan's 1986 holiday message to the people of the Soviet Union was jammed.

"We are told that your radio station is more dangerous than a whole squadron of bombers. As you see, they think highly of you and call you radio bandits and traitors. Of course they forbid us to listen to your broadcasts and since they forbid it, it means that you broadcast the truth and truth hurts; so on the sly everyone listens. The trouble is, the reception is very poor. When you get to the most interesting point, the sound disappears. God willing, we will meet in FREE UKRAINE."

LISTENER IN UKRAINE, SOVIET UNION

★ ★ EIGHT
VOA Listeners

Who listens to the Voice of America? VOA listeners are young and old, male and female, educated and not educated. Using information from surveys conducted by VOA among their listeners, and information from talking to travelers from other countries who know VOA listeners, the typical listener can be identified. That typical listener is a male who has completed secondary school and wants to know about other countries and to learn about politics and government. He either owns a shortwave radio or is able to listen to VOA on a friend's radio. He is so eager to hear the programs that he is willing to put himself in danger to do so if he lives in a country where listening to VOA is forbidden. He also listens to other international radio stations, and compares the coverage and viewpoints of the various stations.

What do 120 million people who tune in to VOA each

year most want to hear? The news. For many listeners, VOA is the only way they can find out about what is happening in the world.

Listeners also like to find out about life in America. Sometimes the only glimpse of America for people in other countries is American television shows. It has been said that many people around the world get their ideas about America from watching "Dallas." But a television show put on for entertainment is not necessarily trying to present a picture of typical life in the United States. VOA gives listeners the kind of information they are eager to receive.

How much does an American worker earn each year? How big are their houses? How do children get their schooling? How much does a meal in a restaurant cost? VOA listeners enjoy hearing information about Americans, and how their lives differ from their own.

Many people dream about America. They picture the buildings, rivers, food, and people, knowing that they will never actually visit the United States. But VOA is the next best thing to visiting America. Listeners can imagine what this country is like through the eyes and voices of the VOA broadcasters.

When a favorite American television newsman retired several years ago, Americans talked of losing a trusted friend. People in other countries feel this same way about VOA broadcasters. Every day, the listeners tune in VOA and welcome the broadcasters into their homes. They learn to recognize their voices, and after listening for several months these broadcasters seem like old friends.

Listeners often write to their VOA friends. VOA received nearly 400,000 letters from listeners in 1985. These letters cost over $200,000 in postage! Listeners not only

74 ★ The Voice of America

A VOA correspondent in New Delhi, India, with listeners

> "I like shortwave because by listening to shortwave I can make an impossible dream come true: to travel around the world. When I listen to the Voice of America, I imagine that I am in Washington going down Independence Avenue and entering the VOA studios where the entire Brazilian Branch is there to give me friendly greetings, who by now, I consider close friends. After visiting the studios, I visit the rest of the city, meeting people and getting to know their culture."
> **LISTENER IN LORENA, BRAZIL**

make comments on VOA programs, but they write about their personal beliefs and lives.

Some letters discuss the political situation in their homeland.

> "I would appreciate a textbook and a cassette. I just don't dare ask for a cassette player because such a request would make me feel ashamed. Yes, ashamed of living in such a beggar country ruled by communists who have brought us to the present state of poverty, misery, and want; a country where you can't even buy a roll of toilet paper."
> **LISTENER IN POLAND**

Other letters express the hope for peace in the world.

76 ★ The Voice of America

> "In spite of difficulties, I listen to your programs every day. I have to warm the batteries of my radio in the fire in order to listen to your dear voices. The programs you transmit are the life of my soul and my greatest hope for a better future."
> **LISTENER IN MOZAMBIQUE, AFRICA**

Eastern Europeans are VOA listeners, despite the difficulties.

Listeners sometimes comment on tragedies that have happened in America. They sympathize with Americans on their losses, which is remarkable when one considers the daily tragedies that occur in the lives of the writers.

> "I convey my deep sympathy over the Challenger's explosion which killed seven astronauts. Those seven women and men do not belong only to the U.S.; they belong to the world. Please convey my condolence to their families. During the past seven years, the Muslim people of Afghanistan have been fighting empty-handed with a super-power to gain their freedom and independence. There is not a single day here that does not bring killing and torturing for the Afghan people, but still we share the pain and unhappiness of the Challenger's tragedy with the people of the U.S."
> **LISTENER IN HERAT, AFGHANISTAN**

All the letters received by VOA are not positive. Occasionally letters are critical of U.S. policies or government. But every letter gets a personal reply. Some letters are received written in languages other than English, and they are answered by people in that particular language service. There is a whole office devoted to answering listeners' mail.

Sometimes the VOA staffer has to answer questions about America. In reply to a letter from Nigeria in Africa: "When people ask, 'Who is the founding father of America?' we refer to George Washington."

Listeners discover that things are done a little differently in America. A writer in Kenya, Africa, asked about American farmers and got this reply: "For many American farmers who must work large tracts of land, the job would be impossible without tractors, threshers, and other mechanized devices. Imagine a farmer with 100 head of cattle having to milk them himself!"

In addition to letters, listeners are sent VOA promotional items if they request them: the VOA calendar, VOA T-shirts, program schedules, books on U.S. government and geography, VOA buttons, rulers, flags, and gym bags.

Voice magazine was introduced in January, 1984. Produced by the Audience Relations Division of VOA, it contains listeners' letters, profiles of broadcasters, articles about America, and information about VOA programs. Lis-

Voice of America promotional materials

teners can request to be put on the mailing list. Circulation exceeds 130,000 copies.

Listeners seem to appreciate the time the VOA staff spends answering mail. A listener in Finland wrote: "Many, many thanks for your letters. I asked other stations the same thing but none of them answered with a personal letter. It seems VOA really wants to help and serve its listeners as individuals."

Sometimes the VOA staff provides information so that a student can complete an assigned school report. At other times, the VOA staff can help make dreams come true. On one of the "Country Music USA" programs, host Judy Massa interviewed Ronnie Milsap, a country music singer. He told her about the foundation he had started to help people who were blind. A Peace Corps volunteer in the West African country of Cameroon heard that VOA program. The volunteer wrote a letter to VOA describing his Peace Corps assignment at a center for persons who are blind. VOA forwarded his letter to Ronnie Milsap, and the Milsap Foundation answered with a $1000 grant.

VOA staff persons occasionally travel to other countries and meet their listeners. Meeting a few fans can be a rewarding experience; Meeting hundreds of members of a VOA Fan Club can be overwhelming.

Stan Schrager, a VOA Division Chief, recently did this in the South Asian country of Bangladesh. Schrager made eighteen appearances on a five-city tour. Hundreds of VOA fans showed up to meet him as he rode in a rickshaw, gave speeches, and laid the marble cornerstone for a new VOA Fan Club building.

VOA is very popular in Bangladesh. The people send nearly 3,000 letters monthly to the VOA Bangla Service

Stan Schrager visits a VOA Fan Club in Bangladesh.

in Washington. The fifteen-to-twenty-five-year-old young men who regularly gather by an oil lamp or single light bulb to listen to VOA form Fan Clubs.

There are some 200 such VOA Fan Clubs. Some members just gather together to listen to the daily 7:00 A.M. and 10:00 P.M. hour-long programs. Others have parties and put together exhibitions about VOA. In one village, the Fan Club performed a community service by buying a rickshaw for a man with disabilities.

Schrager was a celebrity, signing autographs and kissing babies. Receptions like this are what make days spent talking into a microphone in an empty studio worthwhile.

> "Our club does voluntary work and helps others in different jobs: planting trees in our area, cleaning up projects, assisting in education to the poor, working in blood donation camps, and helping with some cultural projects. We meet every second and fourth Sunday to discuss various VOA programs broadcast during those weeks."
>
> LISTENER IN A **VOA FAN CLUB** IN CALCUTTA, INDIA

VOA has started a newsletter for listeners' Fan Clubs. The first issue was dated August, 1986, and contained articles on VOA history and a profile of a famous Bangla novelist and VOA broadcaster, Dilara Hashem.

At times, VOA listeners get to go on the other side of the microphone and become a part of a VOA broadcast.

Dr. Haing Ngor, 1985 winner of an Academy Award for his performance in the movie, *The Killing Fields,* was interviewed on VOA by the Khmer Language Service. Before his escape from Cambodia in 1979, Ngor had been a devoted listener of VOA. He had to listen secretly, though, and used a set of batteries which he carefully recharged and used only for VOA broadcasts. The movie, *The Killing Fields,* focused on the time during the communist takeover of the Southeast Asian country of Cambodia. Ngor's life was similar to much of the action in the movie. He had to flee for his life because he was branded an enemy. He told the VOA broadcaster that *The Killing Fields* was 100 percent true, but it was tame compared to what really happened in his country.

Ngor reached the Cambodian border in May, 1979, but had to wait in hiding for six months before he was able to get into Thailand, a country bordering Cambodia. Cambodians listening to the VOA Khmer Language Service program heard Ngor's interview. Ngor, once a VOA listener, got to talk to his fellow Cambodians through the Voice of America.

Another VOA listener recently went on the air. A nun from Bangladesh took the VOA tour during a visit to the United States. In talking to the VOA tour guide, she discovered that a going-away party was planned for that very afternoon for VOA broadcaster Pat Gates, the popular weekend hostess of the VOA morning program. The nun had listened to Pat Gates for the past ten years, and was eager to meet her. So she accompanied the VOA tour guide to the party, where VOA staffers had gathered to wish Gates well in her new position as U.S. Ambassador to Madagascar. The nun not only got to meet Gates, she took the microphone at the party and made a heartwarming speech about listening to Gates back in Bangladesh that brought tears to the eyes of many people present. Afterward, she was interviewed by the Bangla Language Service for broadcast to Bangladesh.

Listeners are very important to a radio station. After all, without listeners what is the point of broadcasting? VOA listeners may be harder to contact because they are so far away, but they are still an essential part of VOA.

"VOA has grown into a giant, winning worldwide attention and respect from thousands of listeners in many countries in many parts of the globe. I receive your broadcast in Guyana loud and clear, day and night, and I love it all."
   LISTENER IN NORTH RAIMVELDT, GUYANA

★ ★ NINE ★ ★ ★ ★ ★ ★ ★ ★ ★

Broadcasters

Habib (not his real name) led a very quiet and happy life with his family in Afghanistan. He was a regular VOA listener, especially enjoying the early morning "Breakfast Show."

Then the Soviets invaded his country, and changed his life completely. As a high-ranking government employee, Habib had to actively support the policies of the new communist government. But Habib did not support these policies, and he knew he could not keep lying about the way he felt toward the communist government. So he didn't have a choice. He had to escape and take his family with him.

Escape meant fleeing across the border to Pakistan, taking the tremendous risk of being shot and killed. It involved at least three days and nights of travel, facing dangers of starvation, frostbite, and disease. The first time he tried to escape, he was found and jailed. But the second try was successful, and Habib and his family left their homeland for a new life.

Today, Habib is a broadcaster for the Dari Language Service of the Voice of America. He broadcasts programs to his homeland of Afghanistan.

For some people, a job is merely a place to go for eight hours a day to earn money to pay bills. For others, like Habib, their job means much more. "I'm glad that by joining VOA as an international broadcaster, I have been afforded an opportunity to indirectly contribute toward the cause of our people who are fighting to safeguard their freedom. I feel I have a constant spiritual link with our people, and it gives me a sense of satisfaction to serve them through such a channel."

Many of the VOA broadcasters originally lived in another country. They have many dramatic stories to tell about the conditions that made them leave their country and come to the United States. But some broadcasters are native-born Americans who studied radio and a foreign language at college. Some grew up in a home where two languages were spoken, so they are fluent in Spanish, French, or Vietnamese.

Not everyone who applies for a job as a VOA broadcaster is hired. The job description outlines the education and experience you need. There are tests you have to pass. You have to be fluent in the language of the language service where you have applied. You have to be able to translate from that language into English, and from English into that language. Finally, you have to take a voice test, to determine whether your voice would sound good on the radio.

86 ★ The Voice of America

Two Chinese broadcasters

Imagine listening to a radio station. The announcer is speaking English, but has such a thick French accent you can barely understand what is being said. Would you continue listening to that station, or would you start switching stations to find one that you could understand more easily?

The people in Saudi Arabia want to listen to a radio station that uses clear, understandable Arabic. The people in the Dominican Republic want to hear clear Spanish. Everyone wants to hear his language spoken as they speak it, without an American accent. So it is very important for the broadcasters to speak the language well.

> "Every time I turn on my radio, I look for VOA because it is very easy to understand. The accent and diction of the broadcasters are so clear and so nice to hear as compared with other worldwide radio stations."
> **LISTENER IN SAUDI ARABIA**

Some broadcasters come to VOA with broadcasting experience. They have worked at a radio station before, either in the U.S. or in their home country. For other broadcasters, VOA is their first radio job.

The broadcasters at VOA have the added responsibility of being sensitive to the culture of the country receiving their broadcasts. In deciding the programs to be aired, they have to consider that topics and issues that are perfectly acceptable in the United States may not be appropriate for another country. For example, a feature on successful hog-raising by a 4-H Club at a state fair would be a nice program for a commercial radio station in the United States. Sounds of the fair, including oinks and snorts, would make the program even more interesting. This same program would offend listeners in a Moslem country, since they consider pork unclean. Also, broadcasters do not want their listeners to feel that Americans are better than they are. The differences between the countries may be pointed out in a program, but broadcasters must be aware of not hurting the feelings of the listeners by excessive bragging about the United States.

The Senior Managing Editor in the Pashto Language Service had a long, distinguished career in the Afghan army before he came to America and VOA. Jabir (not his real name) came from an intellectual family; his father was a professor at a local university in Afghanistan. When he was a teenager, Jabir was selected by the government to attend the military academy. That was quite an honor in those days, and so he went to school and began his career in the army. Jabir soon became an expert on international military affairs. He wrote on this subject for magazines, and even had his own column. Concerned that the people in Afghanistan needed a written record of their military history, he wrote a three-volume set that was published from 1964 to 1972. During this period, Jabir traveled to various countries around the world, continuing his education and lecturing on war and diplomacy.

Jabir was a career military man, a world traveler, and a nationally known writer. He lived with his family in the country where he was born. But he was not to live out his life in this way. He was forbidden by the government to write, and Jabir essentially began the second half of his life.

"You are lost. You say the government way, or you don't say it. You do the government way, or you don't do it. You write the government way, or you don't write."

When the communists invaded Afghanistan, it affected everyone, including his son, who was nine at the time. The boy gathered with some of his young friends one day and began throwing stones at a Soviet Army man. Jabir looked everywhere for his son that afternoon, and finally found

An interview on the Bangla Language Service program

him at the police station where he'd been taken after the stoning incident. Jabir had to write on a paper that his son would never again stone a Soviet military man before he was allowed to take the child home. A year later, the boy, his six-year-old sister, his mother, and Jabir were walking across the mountains on their way to Pakistan.

Their trek took ten days, but the preparation had taken much more time. The villages outside the major cities were liberated. People escaping from the cities journeyed from farm to farm in these outlying areas, but the trip had to be well planned. Jabir had made contact with people in these villages weeks before he planned to leave, and made all the arrangements. Secrecy was very important. If it was discovered that you planned to leave, you would be put in jail. Jabir knew that he had to make it seem as if he would still be in his home the next day and week. Some people put their cars in garages to be fixed, giving the impression that they would be there to pick up their cars once they were repaired. Jabir's family left clothes hanging on their line. Of course, they were not there to bring the clothes in the next day.

The traveling was done at night, so the helicopters could not spot them. Jabir's children were tired of walking, and they kept asking questions. The children were carried when they were too exhausted to continue, but the questions were not answered until they were safe in Pakistan.

Once there, Jabir joined the military committee of the resistance. In 1982, Jabir and his family came to the United States. The Americans who were helping them settle had heard that VOA was recruiting people to work in the new Pashto Language Service. When Jabir arrived in America, there was a government employment application waiting for him. His second life began with a new career at Voice of America.

"I feel great. They took my pen away from me, because I was a writer. They took the microphone away from me. I hoped someday to find it. I've found it now at VOA."

Memories of his former life remain with Jabir, and his past experiences and studies are reflected in some of the Pashto programming. Jabir worked on a radio program commemorating the 150th anniversary of the American battle of the Alamo in Texas. He compared that famous battle with an Afghan battle that was similar in many ways. Jabir also helped prepare two award-winning programs for the Pashto Language Service. The programs, "Wars and Heroes," and "The Sixth Anniversary of the Russian Invasion of Afghanistan" were recipients of the 1985 VOA Annual Excellence in Programming Awards.

Let's follow a broadcaster into a VOA studio to discover what happens. We'll call him Eduardo, a member of the staff of the Portuguese to Africa Language Service.

★ Eduardo has already worked with the rest of the broadcast team in preparation for the day's broadcast. He is familiar with the programs for the day, and he knows the order of the different shows. A rehearsal was done earlier in the day so that Eduardo would know exactly what to do.

★ Eduardo goes to the studio he will use for that day's broadcast. A green "In Use" light on the studio door means that the engineer is in the control room, the power is on, and they are ready to begin broadcasting. Once the actual broadcast begins, a red "On Air" light will go on.

★ The door is heavy, because it helps block the noise of the hallway from the studio and control rooms. Eduardo enters the studio and sits at a table. A microphone is right in front of him. The broadcast will start in 15 minutes, and Eduardo uses this time to have a last-minute rehearsal.

★ Through a window on the far wall of the soundproof studio Eduardo can see the engineer and producer in the control room. The control room contains tape machines, turntables for records, and a control board with many

A VOA engineer in the control room

switches and levers, which the engineer uses to mix sounds together, adjust the volume of the broadcaster's voice, and send the sounds to the Master Control Room from which they are sent abroad.

The producer speaks Portuguese, and since Eduardo is wearing a headset, she can speak to him if the need arises. But hand signals are also used. Eduardo can see the producer's hand signals through the window.

★ Eduardo does a microphone check, talking into the microphone so that the engineer can adjust the sound levels on the control board. He adjusts the microphone, and looks over his papers to make sure they are in the correct order.

★ He glances at the large clock in the studio, and notes that his show will start in ten seconds. He watches the producer as she gives him the "Stand By" hand signal, and then "You're On."

The producer gives the broadcaster the signal, "You're on!"

★ The broadcast begins. Eduardo starts reading the news items. Several minutes later, another staff person from his service enters the studio and hands him another news item. Eduardo hears the producer in his headset telling him to substitute the new item in place of the next page he has in his hands. He continues to read the news as this is going on. Without stopping or slowing down he substitutes the latest news item, and the listeners have no idea that a last-minute change has taken place.

★ He continues reading his scripts, dividing his attention between the clock, the producer, and the paper he is reading. Always, Eduardo remembers to speak clearly.

Broadcasters do not spend their entire workday talking into a microphone. Some language services only broadcast for one hour a day. The rest of the time is spent preparing for the show, and planning future shows.

Some programs are taped, for broadcast at a later time. For most shows the engineers supply musical interludes and various sound effects, so that the programs are not entirely just voices. VOA has several programs that include guests, and the broadcaster is interviewing or talking to another person. Many programs use more than one broadcaster, so that more than one voice is heard. There has to be a lot of cooperation between the various people in the services to make sure good programming is being aired.

Salman M. Hilmy, the 1985 Outstanding VOA Employee of the Year, is quick to note that credit for his Arabic Branch's success must be shared. "I'm proud of the fact the entire Arabic Branch pulls together. There is no such thing as an individual going it alone in radio anymore."

Each hour of programming is carefully planned. Many people at VOA use the program clock. It is a guide for the broadcasters for filling up their time on the air. There is no rule that you have to use the program clock, but it is strongly encouraged.

The program clock is a set of two circles, one inside the other. They are divided into 60 one-minute segments.

The opening of the broadcast includes an announcement in English, "This is the Voice of America." The broadcaster then identifies what language the broadcast will be using. Then the tune "Yankee Doodle" is played.

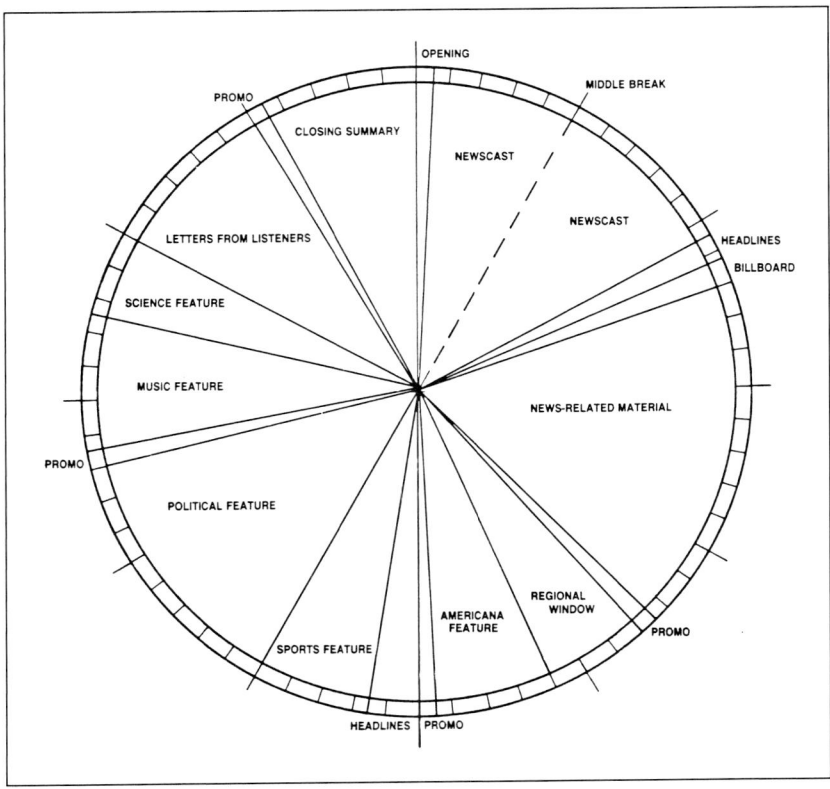

VOA's program clock is a guide for broadcasters.

This gives the listeners time to adjust their radios before the program starts. "Yankee Doodle" slowly fades as the broadcaster starts reading the news.

The newscast usually lasts about ten minutes. In the middle of the newscast, the broadcaster again says that the listeners are tuned in to Voice of America. This is so that people who tuned in after the beginning of the newscast are told they are listening to VOA.

After the newscast ends, the headlines are read, brief descriptions of the top three news stories. The billboard gives highlights about the rest of the broadcast, so that listeners will know what is coming up. Several times during the hour, promos are done. A promo is a promotional announcement, similar to those on television for upcoming shows. A typical promo would be: "On tomorrow afternoon's show, we will discuss possible medical cures for the common cold."

Next on the program clock is the news-related material. News items are explained, possibly with live reports from news correspondents. The regional window also covers news items, but these are about events and people in the listeners' region of the world. The Americana feature is usually a story about life in the United States. Stories from the VOA Voyager van are aired here.

Sports news is next—perhaps an American sports event such as the Super Bowl, or an international soccer meet. The political feature includes comments by people about the news. Sometimes these are programs taped before broadcast time, such as the program "Close-up" produced by the Current Affairs Division. Science features may be topics such as medicine or space travel. Then several minutes are spent reading letters from listeners. The listeners enjoy hearing what other people think of the VOA programs.

At the end of the hour, the broadcaster thanks everyone for listening, and gives a quick word about upcoming programs. Sometimes, a broadcaster will announce the frequency, or number on the radio, on which the program is broadcast. Then "Yankee Doodle" is played again to end that hour of broadcasting.

Charting Voice of America programs

VOA broadcasters are kept very busy, but they enjoy what they are doing. "I passionately love my profession," says a Hindi Language Service broadcaster.

How many other jobs enable you to put joy in someone's day, bring much-needed information to people starved for news, or help bring a bit of peace to this world of ours?

> **"Listening to Western radio is an indirect way of contributing to world peace. By listening to each other, nations can understand each other better."**
> **LISTENER IN THE SOVIET UNION**

★ ★ TEN ★ ★ ★ ★ ★ ★ ★ ★
Radio Marti

Americans often take the availability of information for granted. If you wanted to find out how your favorite hockey team fared in their game the night before, you could buy a newspaper and find out. If you wanted to find out what the President said in his speech the previous day, you could turn on your radio and listen to a news report giving all the details about the speech. If you wanted to enjoy a lighthearted soap opera, you could tune in your television and watch the latest scandals and love stories.

This is not the case in many countries around the world. In the island country of Cuba in the Caribbean, the government controls everything: the newspapers, the radio and television stations, the mails and the telephones. The people only buy, hear, watch, and read things that have been approved by their government.

Communication between Cubans and their families and

friends living in other countries is often very difficult. The government makes it hard for people to stay in touch. It is often difficult to get a call through to the island, and people fear that the mail service is unreliable.

This communications problem made one man desperate. He had come to America from Cuba, leaving members of his family behind. He had not heard from his son for over a year, and was growing very concerned because the boy was approaching his fifteenth birthday. In Cuba, fifteen-year-old boys are eligible for military service and can be drafted into the army at any time. The father did not know where his son was, and he worried that the boy would be drafted and sent to a country overseas, and never be heard from again. The man took a chance by calling a toll-free telephone number he had seen advertised in a Spanish language newspaper.

He reached a person in Washington, D.C., who worked with the radio program *"Puente Familiar,"* meaning "Family Bridge" in Spanish. The man taped a message to his son, appealing for some news about the boy's welfare and whereabouts. This message was broadcast to Cuba over Radio Marti, a part of the Voice of America.

Today, the man is very happy. Four weeks after his message was broadcast he received a letter from his cousin in Cuba, who wrote that a neighbor had listened to Radio Marti and heard the message. The father learned that his son was safe, in school, and that he would write to his father soon.

Radio Marti is a semi-independent organization within VOA, created by an act of Congress and signed into law by President Reagan in 1983. This Cuban Service is administered separately from other VOA functions, and its head reports directly to the USIA Director and to USIA's

Associate Director for Broadcasting, the VOA Director. Its operations are housed in a building in Washington, D.C., a block away from VOA broadcasting headquarters. Radio Marti broadcasts over 16 hours of programming every day. It has been on the air since May 20, 1985, at 5:30 A.M.

"Family Bridge" has become a very popular radio program. People call and tape messages about all kinds of things, hoping their friends and families in Cuba will be listening. Over twenty-five calls a day are received, most of them variations on "Please write, Aunt Louisa" or "Grandpa, our son was born last week and we named him after you." Cubans have written their families in the United States and requested them to call "Family Bridge" and tape a message so that they can listen to their voices.

"Family Bridge" has increased from its original one-

A Radio Marti broadcaster

hour weekly program because of the backlog of messages that needed to be broadcast. Radio Marti has many other programs and features, but this particular show is a real service for listeners in Cuba.

> "The last thing a woman I know told me before I left Cuba was that if ever I had the opportunity to talk to a reporter or an employee from Radio Marti to tell them with all assurances that the people of Cuba listen to Radio Marti."
>
> **FORMER LISTENER NOW IN THE U.S.**

Although Radio Marti has not been on the air very long, it has already caused controversy. In 1980, over 120,000 Cubans left the fishing port of Mariel and traveled by boat to the United States. Among these refugees were many Cuban criminals who had been allowed to leave prisons in Cuba. After many negotiations, Cuba finally agreed to take back the criminals, but only a few hundred were returned. According to many observers, when Radio Marti began broadcasting, it bothered Fidel Castro, the leader of Cuba, so much that he suspended the agreement about the criminals returning to Cuba. Radio Marti became a bargaining point between the two nations. Unless the U.S. stopped the broadcasts of Radio Marti, nothing more would be done about the return of Cuban criminals. Radio Marti was not stopped, and the agreement remains suspended.

Radio Marti was named after José Julian Marti, a Cuban hero of the nineteenth century who dedicated his life to freeing Cuba from Spanish rule. He once said, "We love freedom because in it we see the truth."

On the occasion of Radio Marti's first broadcasts, Congressman Dante B. Fascell (D-Florida) acknowledged the connection between José Marti and his namesake, Radio Marti. "Freedom to think and speak, safe from reprisals by governments, political parties, or any who would impose their will, that is what Marti fought for and that is the goal and meaning of Radio Marti."

News reports take up almost half of Radio Marti's broadcast day. The news reports often differ from the news heard on the official Cuban radio stations. But, according to Cubans who have recently moved to the U.S., Radio Marti has good credibility with the Cuban people. They believe that the news on Radio Marti is accurate.

> **"Radio Marti's popularity spread until everybody was listening to it. In the municipality in Havana province where I lived, the whole town listened to it."**
> **FORMER LISTENER NOW IN THE U.S.**

Radio Marti's special programs and features are very popular. "Esmeralda," a 274-chapter radio soap opera, recently ended. It was a simple love story that appealed to listeners of all ages and interests. The happy ending, where true love triumphed, proved that love can overcome all obstacles. "Esmeralda" 's story provided an escape for its listeners.

A *New York Times* report from Havana stated, "Mention the name Esmeralda to many people in Havana and you're likely to get a smile and a conspiratorial wink. It seems a growing number of Cubans are following the travails of Esmeralda. In the first day or two of the broadcasts it

A Radio Marti engineer

became difficult to find anyone who would acknowledge they listened. Many Cubans seemed nervous when foreign journalists asked questions about Radio Marti. Now, ordinary Cubans seem more relaxed about 'RM' as some are calling it, although most tell strangers they are not regular listeners. Nearly everyone, however, seems to know who Esmeralda is."

> "I go to work at 11:30 A.M. and 'Esmeralda' continues until noon. So I listen to the first part, but I have an eighteen-year-old co-worker that lives close to work and she has time to hear the complete story. So she tells me the second half."
>
> **LISTENER IN CUBA**

Radio Marti broadcasts American music, which is enjoyed by young Cubans. This light programming contrasts with much of the Cuban television and radio offerings, and people tape the Radio Marti programs so that they can listen again and again. Possibly because of the popularity of these programs, Cuba is now expanding and improving its own television and radio stations. In trying to compete with the varied programming on Radio Marti, a lot of Western movies have been added to Cuban television, a soap opera was added to one of the Cuban radio stations, and a new superpower radio station went on the air in November, 1985.

> **"Don't think because you don't receive letters from many Cubans it is because we are not listening to the radio station or because we don't like it. Everybody likes the music and the programs you offer to us. We don't have freedom and that makes everybody afraid to write you because of the potential consequences."**
>
> **LISTENER IN CUBA**

Other programs are "Testimony," which centers about an event or a person from Cuba's history, and "Roundtable," which features a discussion on a current topic by a panel of people with different views. One "Roundtable" program had a panel of Cuban children who had recently moved to the U.S. They discussed their feelings about the holiday of Christmas. Cubans are not allowed to celebrate Christmas or to have a Christmas tree. Instead, they get presents on "Kids Day," the first weekend in July that is closest to the anniversary of the revolution.

One of the purposes of Radio Marti, as noted in the law creating the service, is "to promote the cause of freedom in Cuba." One way of doing that is to let the Cuban people know what life is like in the United States. Broadcasters do not have to read a long list of freedoms enjoyed by Americans for the Cubans to realize what freedom means and to appreciate the differences between the two countries. Everyday happenings in the U.S. tell their own story.

When a Cuban immigrant is interviewed on Radio Marti about his steady job and the house he recently purchased here, that's America.

When Radio Marti quotes the Mayor of Miami, Florida, a man who was born in Cuba and now governs a major city in the United States, that's America.

When "Roundtable" is aired on Radio Marti and different points of view are expressed, that's America.

As Senator Edward M. Kennedy (D-Massachusetts) said on one of Radio Marti's first broadcasts, "As the great poet, patriot, philosopher, and liberator, José Marti, understood so well, knowledge is power. And it is fitting that this new station which proudly bears his name is now bringing power through knowledge to the Cuban people."

★ ★ELEVEN★ ★ ★ ★ ★ ★ ★ ★

VOA and You

You don't live in a foreign country and you don't have a shortwave radio, so you cannot hear the programs broadcast by the Voice of America. You never plan to tour the VOA headquarters in the nation's capital. You do not intend to work in radio when you grow up, so you will not become a VOA broadcaster. What does VOA offer you? How does it affect your life?

There are many facts you learn in school that do not seem to have anything to do with your life today, but that will come in handy in the years to come. This is true of VOA. The Voice of America is an important function of the U.S. government. Finding out about it and what it does helps you become more knowledgeable about other countries and the other people who share this world with you.

A VOA broadcaster in the Far East Division says this

about her job: "It is the chance to brush shoulders with all kinds of people, talking to them, asking about their clothing, their customs, and learning and working side by side with them. It is like a large family made out of different kinds of people. It is wonderful . . . Human beings are the same everywhere, we are only different on the surface. Deep down we are universal."

In the United States you can listen to the radio stations of your choice, and watch the television programs you want to see. The U.S. government does not restrict what you see and hear. In learning about VOA you discover this is not the case in all the areas of the world. In Iran, rock music is banned. In Cuba, the news is censored by the government. A Farsi Language Service broadcaster described the difference between the children in America and the children in Iran as "the difference between being free and not being free."

Visitors to Voice of America are welcomed in many languages.

A broadcaster from Afghanistan feels that VOA has a message for children. "I want the American children to know that the Afghan children are in desperate need of their moral and material support. I think they should raise their voices in support of Afghan children who are in danger of extermination."

Even though you cannot listen directly to VOA, you can keep up with what is going on in the world. Read a newspaper or news magazine. Pay attention to the news on television. Take advantage of your freedom to get accurate and truthful news. Know what goes on in other countries.

VOA broadcasts programs to make people outside this country aware of the many different cultures and differences in America. Begin to notice the differences—different accents, different ways of doing things, different expressions. A sandwich made with a bread roll filled with meat and cheese is called a sub, a hoagie, a hero, or a grinder, depending on what area of America you are in.

Sometimes we tend to overlook interesting places in our own communities. If the VOA Voyager van were visiting your hometown, where would you suggest it go? What is there about your town that you feel people overseas would like to know? You represent America. What would you want to tell a foreigner about the way you live?

The Voice of America, celebrating its 45th anniversary in 1987, was created in order to provide an objective source of information about this country for the rest of the world. Today it continues to broadcast information to all parts of the world.

VOA is not universally loved. Some people accuse it of broadcasting propaganda, others complain that it represents only part of the total American story. But millions

VOA provides information about America to listeners through interviews with newsmakers such as Sally Ride, first American woman astronaut.

of people continue listening to VOA because they want the information it provides.

Richard Carlson, Director of VOA, says: "Anyone who listens to VOA knows that we broadcast warts and all. The news that we put forth is as unbiased, and more so in some cases, as other commercial broadcasters. There is no American network that doesn't have criticism. I think we have less than the commercial broadcasters do.... There is no interference with the news operation ever by the State Department or the political administration. None. Our news is truthful and honest."

VOA listeners feel that the Voice of America is an essential part of their lives.

> "Since we don't have daily newspapers here, the radio is the only link to the rest of the world. We hear lots of different voices. I have chosen to listen to yours. For it's the voice of a great, freedom-loving country. It's an honest voice that doesn't shut up others. It permits criticism and the expression of different opinions. The information you provide us with is clear and abundant. The programs are as varied as they are interesting. The commentaries are long enough to allow the listener to understand and to remember. All this may seem like flattery to you, but it's only an expression of my deep admiration of the Voice of America."
>
> **LISTENER IN BURKINA FASO, AFRICA**

VOA Languages (AS OF AUGUST, 1986)

★ Office of News and English Broadcasts (NEB)
  English broadcasts are directed to every area of the world.
  NEB includes the News Division, Current Affairs Division, English Programs Division

★ African Division
  Amharic to Ethiopia
  English to Africa
  French to Africa
  Hausa to Ethiopia
  Portuguese to Africa
  Swahili to East Africa

★ American Republics Division
  Portuguese to Brazil
  Spanish to Latin America

★ East Asia and Pacific Division
  Burmese to Burma
  Chinese to China
  Indonesian to Indonesia
  Korean to Korea
  Khmer to Cambodia
  Lao to Laos
  Thai to Thailand
  Vietnamese to Vietnam

★ European Division (to eastern Europe)
　　Albanian to Albania
　　Bulgarian to Bulgaria
　　Czech/Slovak to Czechoslovakia
　　Estonian to Estonia
　　Greek to Greece
　　Hungarian to Hungary
　　Latvian to Latvia
　　Lithuanian to Lithuania
　　Polish to Poland
　　Portuguese to Portugal
　　Romanian to Romania
　　Slovene and Serbo-Croatian to Yugoslavia
　　Spanish to Spain
　　Turkish to Turkey

★ Near East, North Africa, and South Asia Division
　　Arabic to countries in the Middle East and
　　　North Africa
　　Farsi (Persian) to Iran
　　Dari and Pashto to Afghanistan
　　Bangla to Bangladesh and the West Bengal state
　　　of India
　　Hindi to India
　　Urdu to Pakistan

★ USSR (Soviet Union) Division
　　Armenian to the Armenian Soviet Socialist Republic
　　Azerbaijani to the Azerbaijan SSR
　　Georgian to the Georgian SSR
　　Russian to Russian Soviet Federated Socialist Republic
　　Ukrainian to the Ukrainian SSR
　　Uzbek to the Uzbek SSR, China, Afghanistan

★ ★ ★ ★ ★ ★ ★ ★ ★ ★ ★ ★ ★

VOA Timeline

1940	Coordinator of Inter-American Affairs (CIAA), formed, begins radio broadcasts to Latin American countries
July, 1941	Coordinator of Information office (COI) formed, containing the Foreign Information Service (FIS)
February 24, 1942	First VOA broadcast out of the FIS studios in New York City
June, 1942	Office of War Information formed, from several agencies, including COI and FIS
1942	Construction begins on first VOA transmitters
August, 1945	VOA is moved into the State Department
January, 1948	Smith-Mundt Act (Public Law 402) is passed. Provided authority for VOA to broadcast to the world, and prohibited VOA from disseminating information to people in the United States
February, 1948	Soviet Union jams VOA for the first time

October, 1948	VOA produces all of its own programming for first time
June, 1953	United States Information Agency (USIA) becomes an independent government agency reporting directly to the President. VOA is part of USIA.
September, 1954	VOA headquarters moved from New York City to Washington, D.C.
1960	First VOA Charter written and approved by the USIA Director, but not passed into law
July 12, 1976	VOA Charter is passed by Congress and signed by President Gerald Ford, becoming Public Law 94-350
1983	VOA begins modernization program
May 20, 1985	Radio Marti begins
1985	VOA receives almost 400,000 letters from listeners
August 16, 1985	The Zorinsky amendment to the Foreign Relations Authorization Act of fiscal year 1986/87 (Public Law 99-93) reinforces control over VOA information going to Americans
October 15, 1985	VOA targets programming to western Europe for the first time since 1960
December, 1986	VOA has more than 100 radio transmitters, 300 antenna groups, and communication satellites

Index

Afghanistan
 broadcasts to, 48, 60, 62–63, 84–85
 letter from listener in, 77
Africa, broadcasts to, 26–31
Aim, VOA. *See* Objective, VOA
Arabic Branch, 94
Armstrong, Louis, 34
Associated Press (AP), 42

Bangladesh, 28, 79, 82
BBC. *See* British Broadcasting Corporation
Bengali Language Service, 19, 79, 82
Bethany, Ohio, transmitter in, 5
Brazil, letter from listener in, 75
British Broadcasting Corporation (BBC), 13
Broadcasters, VOA, 64–98
Burkina Faso, Africa, letter from listener in, 111

Cambodia, 81–82
Cameroon, letters from listeners in, 29, 79
Carlson, Richard W., 111
 foreword by, vii–ix
Castro, Fidel, 102
CBS. *See* Columbia Broadcasting System
Charter, VOA, 8–9, 115
Children's programming, 59–65
China
 broadcasts to, 46
 letters from listeners in, 48, 53
Clock, program, 95
"Cold war," 6

Columbia Broadcasting System (CBS), 2, 6
Conover, Willis, 16, 31–35
Conover, Mrs. Willis (Evelyn), 35
Coordinator of Information office (COI), 2, 4, 114
Coordinator of Inter-American Affairs (CIAA), 2, 4, 5, 114
Correspondents' bureaus, 45
"Country Music USA" program, 79
Cuba
 broadcasts to, 10, 16, 99–106
 letters from listeners in, 104, 105
Czechoslovakia
 broadcasts to, 16
 letters from listeners in, 11, 65

Dari Language Service, 60, 85
Delano, California, transmitter in, 5
Dixon, California, transmitter in, 5
Dominican Republic, letter from listener in, 37

Editorial, VOA, 10
Eisenhower, Dwight D., 8
Ellington, Duke, 31, 34
"Esmeralda" (soap opera on Radio Marti), 103–104

"Family Bridge" program (Radio Marti), 100–102
Fan Clubs, VOA, 79–81
Farsi Language Service, 23, 25, 36, 37, 64
Fascell, Dante B., 103
Feature programs, 36–50, 54, 59, 66, 68, 69

Fergis Falls, Minnesota, Voyager van visit to, 48
Finland, letter from listener in, 79
"Flowers of Life" (children's program), 64
Ford, Gerald, 115
Foreign Information Service (FIS), 2, 3, 4, 114
Foreign Relations Authorization Act, Zorinsky amendment to the, 10, 115
Foreign Service officers, VOA and, 6, 14

Gates, Pat, 82
Germany, broadcasts to, 3–4, 16
Ghana, letter from listener in, 10
Greece, broadcasts to, 12
Guinea, 29, 30
Guyana, letter from listener in, 83

Hale, William Harlan, 3–4
Hashen, Dilara, 81
Headquarters, VOA, 1, 7, 13, 18–19, 115
Hilmy, Salman M., 94
Hindi Language Service, 8, 19–20
History programs, 64, 65

India
 broadcasts to, 8, 19–20
 letter from listener in, 81
International Phonetic Alphabet, 42
Iran
 broadcasts to, 23–24, 36–37, 64
 letters from listeners in, 23, 25–26

Jackson, Michael, 25
Jamming of VOA broadcasts, 6, 16–18, 68, 70, 114
Japan, broadcasts to, 16
Jazz, 31–35
Jurey, Philomena, 38–42, 45–47

Kavala, Greece, transmitter in, 20
Kennedy, Edward M., 106

Kenya, letter from listener in, 78
Khmer Language Service, 81–82
Killing Fields, The (movie), 81–82

Languages, VOA, 112–113
Latin America, broadcasts to, 2, 4, 114
Latvia, letter from listener in, 67
Liberia, 29, 30
Listeners, VOA, 72–85

Marti, José Julian, 102–103, 106
Massa, Judy, 21, 79
Milsap, Ronnie, 79
Milsap Foundation, 79
Mission, VOA. *See* Objective, VOA
Mozambique, letter from listener in, 76
Mundt, Karl, 6
Murrow, Edward R., 29
Music programs, 21–35, 64, 69, 79, 105
"Music Time in Africa," 26–31
"Music USA," 31–35

National Broadcasting Company (NBC), 2, 6
News programs, 36–50, 54, 59, 66, 69, 73, 103
Ngor, Dr. Haing, 81–82
Nigeria, letters from listeners in, 53, 77

Objective, VOA, 4, 10
Office of War Information (OWI), 4, 5, 114

Pashto Language Service, 88, 90–91
Pearl Harbor, 3
Poland
 broadcasts to, 16, 31, 34
 letter from listener in, 75
Program clock, 95
Program guide, availability of, 12
Promos, 96
"*Puente Familiar*" program (Radio Marti), 100

Radio Afghanistan, 60
Radio Marti, 10, 99–106, 115
Radio Moscow, 13
Reagan, Ronald, 38, 40, 42, 46, 100
Rochelle, Rita, 26, 29–30
Roosevelt, Franklin D., 2, 3
Royce, Bill, 36
Russia
  broadcasts to, 6, 12, 16–17, 31, 66–71, 114
  letters from listeners in, 34, 35, 66, 68, 98

Sarkisian, Leo, 26–31
Saudi Arabia, letter from listener in, 87
Schrager, Stan, 79, 81
Sharma, Achala, 19–20
Sierra Leone, 26
Smith, H. Alexander, 6
Smith-Mundt Act. *See* United States Information and Educational Exchange Act
"Sound of Music" program, 23–26
Soviet Union. *See* Russia
Special English programs, 51–58
State Department, U.S., 5, 6, 8, 111
Sweden, letter from listener in, 56

Tan, Evelyn D., 34–35
Temple, Texas, Voyager van visit to, 50
Timeline, VOA, 114–115
Transmitters, location of, 5, 20
Trinidad, letter from listener in, 32
Truman, Harry, 5

Ukraine, Soviet, letters from listeners in, 18, 71
United Press International (UPI), 42

United States Information Agency (USIA), 8, 29, 38, 100, 115
United States Information and Educational Exchange Act (Smith-Mundt Act), 6–7, 114

Voice magazine, 78
Voice of America
  Annual Excellence in Programming Awards, 91
  Audience Relations Division, 78
  broadcasters, 84–98
  Charter, 8–9, 115
  children's programming, 59–65
  comparison with commercial stations, 12–20
  editorial, 10
  Fan Clubs, 79–81
  first broadcast, 3–4, 114
  headquarters, 1, 7, 13, 18–19, 115
  improvement of services, 16, 115
  languages, 112–113
  listeners, 72–85
  objective, 4, 10
  political appointees, 14
  principles governing broadcasts, 8–9
  program clock, 95
  Soviet Union Division, 66–71
  Special English Division, 51–58
  Special Events Division, 48
  Timeline, 114–115
  you and the, 107–111
Voyager van, VOA, 2, 47–50, 96

Word Book, VOA, 54, 56
Words and Their Stories, VOA, 54–55

"Yankee Doodle," 4, 95–96, 97

Zorinsky, Edward, 10, 115

WITHDRAWN